Lecture Notes in Physics

359

K. Kirchgässner (Ed.)

Problems Involving Change of Type

Proceedings of a Conference
Held at the University of Stuttgart,
FRG, October 11–14, 1988

Springer-Verlag
Berlin Heidelberg New York London Paris Tokyo Hong Kong

Editor

Klaus Kirchgässner
Mathematisches Institut A, Universität Stuttgart
Pfaffenwaldring 57, D-7000 Stuttgart 80, FRG

ISBN 3-540-52595-5 Springer-Verlag Berlin Heidelberg New York
ISBN 0-387-52595-5 Springer-Verlag New York Berlin Heidelberg

Printed in Germany

Printing: Druckhaus Beltz, Hemsbach/Bergstr.
Bookbindung: J. Schäffer GmbH & Co. KG., Grünstadt
2153/3140-543210 – Printed on acid-free paper

This volume is dedicated to
Professor J. K. Hale
on the occasion of his 60th birthday

Scientific Committee

J.M. Ball, Heriott–Watt University, Edinburgh

K.H. Hoffmann, Universität Augsburg

K. Kirchgässner, Universität Stuttgart

R.J. Knops, Heriott–Watt University, Edinburgh

J. Mallet–Paret, Brown University, Providence

L.A. Peletier, University of Leiden

Preface

The phenomenon of spontaneous change of type is well known in the mathematical description of physical systems which undergo phase transitions. Since most of these problems are notoriously difficult, they have withstood satisfactory mathematical treatment for a long time. The best-known classical examples are given by the steady transonic potential flow of a gas and the flow of fluids exhibiting instantaneous elasticity as studied by Maxwell et. al.

In recent years, renewed interest in such problems has arisen in the mathematical community. Several independent approaches, including the theory of crystal structure, the theory of viscoelasticity, and percolation problems, were responsible for this renewal. In addition, many new mathematical methods have been developed for the analysis of such properties as highly oscillatory behaviour and metastability, which are frequently observed in nonlinear systems and phase–transition problems.

It therefore appeared appropriate to organize an international conference on such problems and discuss and survey the state of the art. This conference took place from 11 October to 14 October 1988 at the University of Stuttgart and was sponsored by the Deutsche Forschungsgemeinschaft, the Volkswagenstiftung, and the European Research Council. It was held in honour of J.K. Hale on the occasion of his 60th birthday. His work on slowly moving manifolds and metastability was well represented at this conference.

These proceedings contain most of the invited lectures. They are ordered, according to their emphasis, into five sections. The first treats variational problems with loss of convexity, regularity, and the connection of each with dynamical systems. The second is devoted to phase–transition problems and their formulation, and the dynamics of interfaces. The third contains problems from viscoelasticity, including a basic paper on the physical foundations. In the fourth, problems are considered where there is a change of type from hyperbolic to elliptic states, in particular problems of shocks and of nonlinear waves. The fifth section treats various aspects of degenerate nonlinear parabolic systems, such as reconstruction of initial traces, free boundaries, and problems of finite propagation speed. The appendix contains additional references which have appeared since the conference.

Stuttgart
February 1990

Klaus Kirchgässner

Contents

4. Hyperbolic-Elliptic Problems

5. Degenerate Parabolic Equations

List of Contributed Lectures

J. v. Below (Tübingen):
Changing of connecting operators for nonlinear parabolic network equations.

H. Berger, G. Warnecke, W. L. Wendland (Stuttgart):
Finite elements for transonic flows.

A. Calsina (Barcelona):
Bifurcation of nonradial solutions of the Navier-Stokes equations in a divergent channel.

F. dell'Isola (Napoli), W. Kosinski (Warszawa):
On a singular surface model in a phase transition problem.

M. Fila (Bratislava):
Boundedness of global solutions of nonlinear diffusion equations.

M. Fiebig-Wittmack (Mainz):
Stationary solutions of a scalar parabolic equation with homogeneous Neumann boundary conditions.

B. Fiedler (Heidelberg):
Complex dynamics of one-dimensional reaction diffusion equations with a non-local term.

M. Grinfeld (Edinburgh):
Some remarks on the Cahn-Hilliard equation.

U. Hornung (München):
Capillary surfaces. Boundaries.

B. Kawohl (Heidelberg):
A family of torsional creep problems

P. Knabner (Augsburg):
Mathematical models for the transport of solutes sorbing porous media.

F. Meynard (Lausanne):
Asymptotic behaviour of solutions in a problem of radially symmetric cavitation in hyperelasticity.

A. Mielke (Stuttgart):
An infinite-dimensional center manifold for a problem of mixed type.

P. Poláčik (Bratislava):
On semilinear parabolic systems addmitting a strong comparison principle.

C. Rocha (Lisboa):
Connection matrices for scalar parabolic equations.

K. Rybakowski (Freiburg):
Bifurcations in diffusive biological systems.

J. Solà-Morales (Barcelona):
Singular perturbations of parabolic equations. The hyperbolic and the elliptic cases.

F. Spirig (Rorschacherberg):
Bifurcation of small periodic solutions.

J. Sprekels (Essen):
Thermomechanical phase transitions with nonconvex free energies of Ginzburg-Landau type.

N. X. Tan (Köln):
Local bifurcation from characteristic values with finite mutiplicity and its applications to partial elliptic differential equations.

A. Vanderbauwhede (Gent):
Subharmonic bifurcation in symmetric systems.

G. Warnecke (Stuttgart):
Admissibility of shock solutions to systems of mixed type.

List of Participants

H.-D. Alber, Universität Stuttgart
H. W. Alt, Universität Bonn
J. Ball, Heriot-Watt University, Edinburgh
J. v. Below, Universität Tübingen
H. Berger, Universität Stuttgart
W.-J. Beyn, Universität Konstanz
J. Brenner, Universität Stuttgart
A. Calsina, Universitat Autonoma de Barcelona
J. Carr, Heriot-Watt University, Edinburgh
L. Cesari, University of Michigan, Ann Arbor
F. Dell'Isola, Università degli Studi, Napoli
E. Di Benedetto, Northwestern University, Evanston
M. Fiebig-Wittmaack, Joh. Gutenberg-Universität, Mainz
B. Fiedler, Universität Heidelberg
M. Fila, Universität Heidelberg
D. Flockerzi, Universität Würzburg
G. Fusco, Università degli Studi di Roma
E. Gekeler, Universität Stuttgart
M. Grinfeld, Heriot-Watt University, Edinburgh
M. E. Gurtin, Carnegie-Mellon University, Pittsburgh
K. P. Hadeler, Universität Tübingen
J. K. Hale, Georgia Institute of Technology, Atlanta
K. Höllig, Universität Stuttgart
K.-H. Hoffmann, Universität Augsburg
G. Hofmann, Universität München
U. Hornung, Universität der Bundeswehr, München
H. Jeggle, Technische Universität Berlin
D. Joseph, University of Minnesota, Minneapolis
H. Kaul, Universität Tübingen
B. Kawohl, Universität Heidelberg
B. L. Keyfitz, University of Houston
D. Kinderlehrer, University of Minnesota, Minneapolis
K. Kirchgässner, Universität Stuttgart
U. Kirchgraber, ETH-Zentrum, Zürich
P. Knabner, Universität Augsburg
R. J. Knops, Heriot-Watt University, Edinburgh
W. Kosinski, Polish Academy of Sciences, Warszawa

D. Kröner, Universität Heidelberg
J. Mallet-Paret, Brown University, Providence
F. Meynard, Ecole Polytechnique Fédérale de Lausanne
A. Mielke, Universität Stuttgart
L. Modica, Università di Pisa
P. de Mottoni, Università di Roma
S. Müller, Heriot-Watt University, Edinburgh
J. Nohel, University of Wisconsin, Madison
L. A. Peletier, University of Leiden
W. Pluschke, Universität Stuttgart
P. Poláčik, Universität Heidelberg
G. Raugel, Ecole Polytechnique, Palaiseau Cédex
M. Reeken, Bergische Universität, Wuppertal
M. Renardy, Virginia Tech., Blacksburg
C. Rocha, Instituto Superior Tecnico, Lisboa
F. Rothe, Universität der Bundeswehr, München
K. Rybakowski, Universität Freiburg
D. Sattinger, University of Minnesota, Minneapolis
J. C. Saut, Université de Paris Sud
B. Scarpellini, Universität Basel
J. Scheurle, Universität Hamburg
M. Slemrod, University of Wisconsin, Madison
J. Sola-Morales, Universitat Autonoma de Barcelona
F. Spirig, ETH Zürich
J. Sprekels, Universität - GHS Essen
N. X. Tan, Universität Köln
A. Vanderbauwhede, Rijksuniversiteit Gent
J. L. Vázquez, Universidad Autonoma de Madrid
H.-O. Walther, Universität München
G. G. Warnecke, Universität Stuttgart
W. L. Wendland, Universität Stuttgart
P. Werner, Universität Stuttgart

1. Variational Problems

Dynamics and Minimizing Sequences

J.M.Ball
Department of Mathematics
Heriot-Watt University
Edinburgh EH14 4AS
Scotland

1 Introduction

What is the relationship between the Second Law of Thermodynamics and the approach to equilibrium of mechanical systems? This deep question has permeated science for over a century, yet is still poorly understood. Particularly obscure is the connection between the way the question is traditionally analysed at different levels of mathematical modelling, for example those of classical and quantum particle mechanics, statistical physics and continuum mechanics.

In this article I make some remarks, and discuss examples, concerning one part of the picture, the justification of variational principles for dynamical systems (especially in infinite dimensions) endowed with a Lyapunov function. For dynamical systems arising from physics the Lyapunov function will typically have a thermodynamic interpretation (entropy, free energy, availability), but its origin will not concern us here. Modern continuum thermomechanics provides such Lyapunov functions for general deforming materials as a consequence of assumed statements of the Second Law such as the Clausius-Duhem inequality (c.f. Coleman & Dill [18], Duhem [20], Ericksen [21], Ball & Knowles [12]). By contrast, statistical physics provides Lyapunov functions only for very special materials (the paradigm being the H-functional for the Boltzmann equation, which models a moderately rarified monatomic gas).

Let $T(t)_{t\geq 0}$ be a dynamical system on some (say, topological) space X. Thus (*i*) $T(0) = identity$, (*ii*) $T(s+t) = T(s)T(t)$ for all $s, t \geq 0$, and (*iii*) the mapping $(t, \varphi) \mapsto T(t)\varphi$ is continuous. We suppose that $T(t)_{t\geq o}$ is endowed with a continuous Lyapunov function $V : X \to \mathbf{R}$, that is $V(T(\bar{t})\varphi)$ is nonincreasing on $[0, \infty)$ for each $\varphi \in X$. (In some situations variations on these assumptions would be appropriate; for example, solutions may not be unique or always globally defined.) The central conjecture is that *if $t_j \to \infty$ then $T(t_j)\varphi$ will be a minimizing sequence for V.* If true, this would give a dynamical justification for the variational principle:

$$\text{Minimize } V. \tag{1.1}$$

What are the obstacles to making this more precise? First, there may exist constants of motion that force the solution $T(t)\varphi$ to remain on some submanifold. These constants of motion must be incorporated as constraints in the variational principle. For example, if the constants of motion are $c_i : X \to \mathbf{R}$, $1 = 1, \ldots, N$, so that

$$c_i(T(t)\varphi) = c_i(\varphi) \qquad \text{for all } t \geq 0,\ i = 1, \ldots, N, \tag{1.2}$$

then the modified variational principle would be

$$\min_{\substack{c_i(\psi)=\alpha_i \\ i=1,\ldots,N}} V(\psi)\,, \tag{1.3}$$

where the α_i are constants. Second, there may be points $\psi \in X$ which are local minimizers (in some sense) but not absolute minimizers of V, so that an appropriate definition of a 'local minimizing sequence' is needed. Third, the conjecture is false for initial data φ belonging to the region of attraction of a rest point that is not a local minimizer of V; such exceptional initial data must somehow be excluded. Fourth, the minimum of V may not be attained, rendering even more problematical a good definition of a local minimizing sequence (c.f. Ball [4]). We are thus searching for a result (applying to a general class of dynamical systems, or to interesting examples) of the type:

Prototheorem *For most initial data φ, and any sequence $t_j \to \infty$, $T(t_j)\varphi$ is a local minimizing sequence for V subject to appropriate constraints.*

The trivial one-dimensional example in Figure 1 illustrates a further difficulty. In the example there are three critical points A,B,C. The Lyapunov

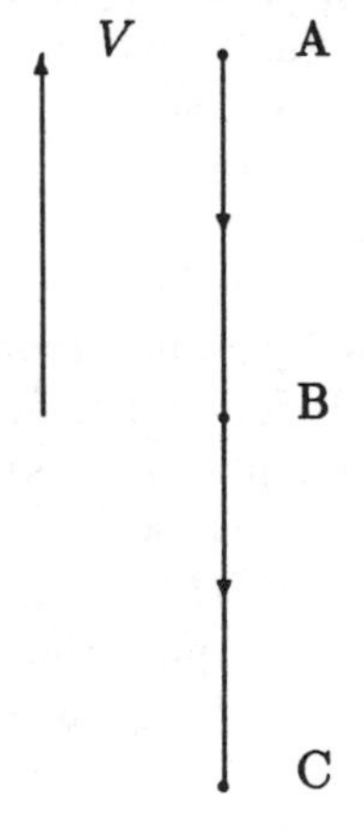

Figure 1:

function V is the vertical coordinate. There is clearly no nontrivial constant of motion, since such a function would have to be constant on the closed intervals

[A,B] and [B,C]. Yet for any $\varphi \in$ [A,B] the solution tends as $t \to \infty$ to a rest point which is not a local minimizer of V. One could have at least three reactions to this example (*i*) that staying in the invariant region [A,B] should be incorporated as a constraint in the variational principle, (*ii*) that the example is not generic, because the rest point B is not hyperbolic, or (*iii*) that stochastic effects should be introduced so that the upper orbit can get through the barrier at B. For, example, taking the point of view (*ii*), a version of the prototheorem can be proved for an ordinary differential equation in $\mathbf{R}^n$.

Theorem 1 *Consider the equation*

$$\dot{x} = f(x), \qquad x \in \mathbf{R}^n, \tag{1.4}$$

where $f : \mathbf{R}^n \to \mathbf{R}^n$ *is* C^1. *Suppose that there exists a continuous Lyapunov function* $V : \mathbf{R}^n \to \mathbf{R}$ *for* (1.4) *satisfying*

$$\lim_{|x| \to \infty} V(x) = \infty, \tag{1.5}$$

and such that if x *is a solution of* (1.4) *with* $V(x(t)) =$*const. for all* $t \geq 0$ *then* x *is a rest point. Suppose further that there are just a finite number of rest points* a_i, $i = 1, \ldots, N$ *of* (1.4), *and that they are each hyperbolic. Then the union of the regions of attraction of the local minimizers of* V *in* $\mathbf{R}^n$ *is open and dense.*

Proof. I sketch the standard argument. By (1.5) each solution $x(t)$ is bounded for $t \geq 0$, so that by the invariance principle (Barbashin & Krasovskii [14], LaSalle [25]) $x(t) \to a_i$ as $t \to \infty$ for some i. Thus

$$\mathbf{R}^n = \bigcup_{i=1}^{N} A(a_i), \tag{1.6}$$

where $A(a_i)$ denotes the region of attraction of a_i. But a hyperbolic rest point a_i is stable if and only if it is a local minimizer of V, while if a_i is unstable then $A(a_i)$ is closed and nowhere dense. □

Note that from (1.6) it follows that under the hypotheses of Theorem 1 there is no nontrivial continuous constant of motion $c : \mathbf{R}^n \to \mathbf{R}$.

Similar results to Theorem 1 can be proved for some classes of (especially semilinear) partial differential equations by combining the invariance principle with linearization (c.f. Hale [23], Henry [24], Dafermos [19], Ball [6,5]), provided the set of rest points is, in an appropriate sense, hyperbolic. However, many interesting examples lie well outside the scope of these results, and no version of the prototheorem of wide applicability is known to me.

The work of Carr & Pego [16] on the Ginzburg-Landau equation with small diffusion shows that, even when the prototheorem holds, solutions may in practice take an extremely long time to approach their asymptotic state, getting stuck along the way in metastable states that are not close to local minimizers.

2 Two variational problems of elasticity

The examples in this section illustrate some of the features described in Section 1. In the first there are nontrivial constants of motion, while in the second the minimum is not attained.

Example 2.1. (*The pure traction problem of thermoelasticity*)
Consider a thermoelastic body in free space, occupying in a reference configuration a bounded domain $\Omega \subset \mathbf{R}^3$. It is assumed that the external body force and volumetric heat supply are zero, that there are no applied surface forces, and that the boundary of the body is insulated. Let $y = y(x,t) \in \mathbf{R}^3$ denote the position at time t of the particle at $x \in \Omega$ in the reference configuration, $v \stackrel{\text{def}}{=} \dot{y}(x,t)$ the velocity, $\epsilon = \epsilon(x,t)$ the internal energy density, and $\rho_R = \rho_R(x)$ the given density in the reference configuration. Then the balance laws of linear momentum, angular momentum and energy imply that

$$\frac{d}{dt}\int_\Omega \rho_R v\, dx = 0, \tag{2.1}$$

$$\frac{d}{dt}\int_\Omega \rho_R y \wedge v\, dx = 0, \tag{2.2}$$

$$\frac{d}{dt}\int_\Omega \rho_R(\epsilon + \frac{1}{2} \mid v \mid^2)\, dx = 0, \tag{2.3}$$

respectively, while as a consequence of the Clausius-Duhem inequality we have that

$$\frac{d}{dt}\int_\Omega -\rho_R \eta(x, Dy, \epsilon)\, dx \leq 0, \tag{2.4}$$

where η denotes the entropy density and Dy the gradient of y. It is assumed that η is frame indifferent, that is

$$\eta(x, RA, \epsilon) = \eta(x, A, \epsilon) \tag{2.5}$$

for all x, A, ϵ and all $R \in SO(3)$. By changing to centre of mass coordinates we may assume that

$$\int_\Omega \rho_R y\, dx = 0, \qquad \int_\Omega \rho_R v\, dx = 0. \tag{2.6}$$

This motivates the variational principle

$$\text{Minimize} \quad \int_\Omega -\rho_R \eta(x, Dy, \epsilon)\, dx \tag{2.7}$$

subject to the constraints

$$\int_{\Omega} \rho_R(\epsilon + \frac{1}{2} \mid v \mid^2)\, dx = \alpha, \tag{2.8}$$

$$\int_{\Omega} \rho_R y\, dx = 0, \qquad \int_{\Omega} \rho_R v\, dx = 0, \tag{2.9}$$

$$\int_{\Omega} \rho_R y \wedge v\, dx = b, \tag{2.10}$$

where $\alpha \in \mathbf{R}$ and $b \in \mathbf{R}^3$ are constant.

The minimization problem (2.7)-(2.10) has recently been studied by Lin [26], who proved that under reasonable polyconvexity and growth conditions on η the minimum is attained at some state $(\overline{y}, \overline{v}, \overline{\epsilon})$. Of course $\overline{y}, \overline{v}, \overline{\epsilon}$ are functions of x alone. As a consequence of (2.5), the minimization problem is invariant to the transformation $(y, v, \epsilon) \mapsto (Ry, Rv, \epsilon)$ for any $R \in SO(3)$ satisfying $Rb = b$. Hence, for any such R, $(R\overline{y}, R\overline{v}, \epsilon)$ is also a minimizer. In fact it is proved in [26] that for any minimizer $(\overline{y}, \overline{v}, \overline{\epsilon})$ there exists a skew matrix Λ such that $\Lambda b = b$, $\overline{v} = \Lambda \overline{y}$, and such that

$$y(x,t) = e^{\Lambda t}\overline{y}(x) \tag{2.11}$$

$$\epsilon(x,t) = \overline{\epsilon}(x) \tag{2.12}$$

is a weak solution of the equations of motion. Furthermore

$$\frac{\partial \eta}{\partial \epsilon}(x, Dy(x,t), \epsilon(x,t)) = \theta^{-1} \tag{2.13}$$

for all t, where θ is a constant. The motion (2.11), (2.12) corresponds to a rigid rotation at constant temperature θ. Note that in this example the Lyapunov function V is constant along nontrivial orbits, such as that given by (2.13). In particular, solutions to the dynamic equations need not tend to a rest point as time $t \to \infty$.

Example 2.2. *(A theory of crystal microstructure)*

Consider an elastic crystal, occupying in a reference configuration a bounded domain $\Omega \subset \mathbf{R}^3$ with sufficiently smooth boundary $\partial\Omega$. Assume that part of the boundary $\partial\Omega_1$ is maintained at a constant temperature θ_0 and at a given deformed position

$$y|_{\partial\Omega} = \overline{y}, \tag{2.14}$$

where $\overline{y} = \overline{y}(\cdot)$, while the remainder of the boundary is insulated and traction free. Then an argument similar to that in Example 2.1, but using a different Lyapunov function, the availability, motivates the variational principle

$$\text{Minimize} \int_{\Omega} W(Dy(x))\, dx \tag{2.15}$$

subject to

$$y|_{\partial\Omega} = \bar{y}, \tag{2.16}$$

where W is the Helmholtz free energy at temperature θ_0 (see Ericksen [21], Ball [3].) It is supposed that W is frame indifferent, i.e.

$$W(RA) = W(A) \tag{2.17}$$

for all A in the domain of W and all $R \in SO(3)$. In addition to (2.17), W has other symmetries arising from the crystal lattice structure, as a consequence of which W is nonelliptic. This lack of ellipticity implies in turn that the minimum in (2.15),(2.16) is in general *not attained* in the natural spaces of admissible mappings. In this case, in order to get closer and closer to the infimum of the energy it is necessary to introduce more and more microstructure. Such microstructure is frequently observed in optical and electron micrographs, where one may see multiple interfaces (occurring, for example, in the form of very fine parallel bands), each corresponding to a jump in Dy. The observed microstructure is not, of course, infinitely fine, as would be predicted by the model here. The conventional explanation for this is that one should incorporate in the energy functional contributions due to interfacial energy; this should predict a limited fineness and impose additional geometric structure (c.f.Parry [29], Fonseca [22]). Since the interfacial energy is very small (witness the large amount of surface observed) it is a reasonable expedient to ignore it, and in fact this successfully predicts many features of the observed microstructure (see Ball & James [10], Chipot & Kinderlehrer [17]). An example in which the nonattainment of a minimum can be rigorously established is the following (a special case of a result of Ball & James [11]). Let $W \geq 0$ with $W(A) = 0$ if and only if $A \in M$, where

$$M = SO(3)S^+ \cup SO(3)S^-, \tag{2.18}$$

where

$$S^{\pm} = \mathbf{1} \pm \delta e_3 \otimes e_1, \tag{2.19}$$

and where $\delta > 0$ and $\{e_1, e_2, e_3\}$ is an orthonormal basis of $\mathbf{R}^3$. Suppose that $\partial\Omega_1 = \partial\Omega$ and that

$$\bar{y}(x) = (\lambda S^+ + (1-\lambda)S^-)\,x, \qquad \lambda \in (0,1). \tag{2.20}$$

Then under some technical hypotheses it is proved in [11] that the infimum of (2.15) subject to (2.16) is zero, and that if $y^{(j)}$ is a minimizing sequence then the Young measure corresponding to $Dy^{(j)}$ is unique and given by

$$\nu_x = \lambda\delta_{S^+} + (1-\lambda)\delta_{S^-}, \qquad \text{for a.e. } x \in \Omega. \tag{2.21}$$

In particular, because ν_x is not a Dirac mass a.e., it follows that the minimum is not attained. The minimizing set M in (2.18) occurs, for example, in the case of an orthorhombic to monoclinic transformation.

It would be very interesting to carry out a dynamical analysis corresponding to the above variational problem, to see if the dynamics produces minimizing sequences with microstructure after the fashion of the prototheorem. This could lead to important insight into a controversial area of metallurgy, that of martensitic nucleation.

3 Some dynamical examples

In this section some infinite-dimensional problems are discussed for which the prototheorem can either be proved or, in the case of Example 3.2, related information obtained.

Example 3.1. (*Stabilization of a rod using the axial force as a control*)
The problem of feedback stabilization of an elastic rod using the axial force as a control leads to the initial-boundary value problem

$$u_{tt} + u_{xxxx} + \left(\int_0^1 u_{xx} u_t dx\right) u_{xx} = 0, \qquad 0 < x < 1, \tag{3.1}$$

$$u = u_{xx} = 0, \qquad x = 0, 1, \tag{3.2}$$

$$u(x,0) = u_0(x), \quad u_t(x,0) = u_1(x), \qquad 0 < x < 1. \tag{3.3}$$

Here $u(x,t)$ denotes the transverse displacement of the rod, while the boundary conditions (3.2) correspond to the case of simply supported ends. This and similar problems were formulated and analyzed in Ball & Slemrod [13]. Using the Lyapunov function

$$V(t) = \int_0^1 \frac{1}{2}(u_t^2 + u_{xx}^2)\, dx, \tag{3.4}$$

which has time derivative

$$\dot{V}(t) = -\left(\int_0^1 u_{xx} u_t \, dx\right)^2, \tag{3.5}$$

it was proved that if $\{u_0, u_1\} \in X \stackrel{\text{def}}{=} (H^2(0,1) \cap H_0^1(0,1)) \times L^2(0,1)$ then the unique weak solution $\{u, u_t\}$ of (3.1)-(3.3) satisfies

$$\{u, u_t\} \rightharpoonup \{0,0\} \qquad \text{weakly in } X \text{ as } t \to \infty. \tag{3.6}$$

Considered as a functional on X, V has only one critical point $\{0,0\}$, which is an absolute minimizer. The conclusion of the prototheorem therefore holds if and only if

$$\{u, u_t\} \to \{0,0\} \qquad \text{strongly in } X \text{ as } t \to \infty. \tag{3.7}$$

This has recently been proved by Müller [28] by means of a delicate analysis of the infinite system of ordinary differential equations satisfied by the coefficients $u_j(t)$ of the Fourier expansion

$$u(x,t) = \sum_{j=1}^{\infty} u_j(t)\sin(j\pi x) \tag{3.8}$$

of a solution. Müller also established the interesting result that given any continuous function $g : [0,\infty) \to (0,\infty)$ with $lim_{t\to\infty} g(t) = 0$ there exists initial data $\{u_0, u_1\} \in X$ such that the solution of (3.1)-(3.3) satisfies

$$V(t) \geq Cg(t) \tag{3.9}$$

for all $t \geq 0$ and some constant $C > 0$. Thus solutions may have an arbitrary slow rate of decay as $t \to \infty$. It is an open question whether strong convergence holds for the case of clamped ends

$$u = u_x = 0 \qquad \text{at } x = 0, 1, \tag{3.10}$$

or for various other feedback stabilization problems for which the analogue of (3.6) was established in [13].

Example 3.2. (*Phase transitions in one-dimensional viscoelasticity*)
Consider one-dimensional motion of a viscoelastic rod. The equation of motion is taken to be

$$u_{tt} = (\sigma(u_x) + u_{xt})_x, \qquad 0 < x < 1, \tag{3.11}$$

with boundary conditions

$$u = 0 \text{ at } x = 0, \qquad \sigma(u_x) + u_{xt} = 0 \text{ at } x = 1, \tag{3.12}$$

and initial conditions

$$u(x,0) = u_0(x), \qquad u_t(x,0) = u_1(x), \qquad 0 < x < 1. \tag{3.13}$$

For simplicity, assume that

$$\sigma(u_x) = W'(u_x), \qquad W(u_x) = (u_x^2 - 1)^2. \tag{3.14}$$

Let

$$V(u,p) = \int_0^1 [\frac{1}{2}p^2 + W(u_x)]\, dx. \tag{3.15}$$

Then $V(u, u_t)$ is a Lyapunov function for (3.11)-(3.13) with time derivative

$$\dot{V}(u, u_t) = -\int_0^1 u_{xt}^2\, dx \leq 0. \tag{3.16}$$

The corresponding variational problem

$$\operatorname*{Min}_{X} V, \tag{3.17}$$

where $X = \{\{u,p\} : u \in W^{1,\infty}(0,1), u(0) = 0, p \in L^2(0,1)\}$ has uncountably many absolute minimizers, given by any pair $\{u,0\} \in X$ with $u_x = \pm 1$ a.e.. In particular it is easily proved that given any smooth function v on $[0,1]$ with $v(0) = 0$ and $| v' | \leq 1$, there exists a sequence $\{u^{(j)}, 0\}$ of absolute minimizers such that $u^{(j)} \overset{*}{\rightharpoonup} v$ in $W^{1,\infty}(0,1)$. This raises the interesting question as to whether a solution $\{u, u_t\}$ to (3.11)-(3.13) could exhibit similar behaviour, converging weakly but not strongly to a pair $\{v, 0\}$ which is not a rest point. This question was resolved by Pego [30], following earlier work of Andrews & Ball [1]. Pego showed that for any solution $\{u, u_t\}$, as $t \to \infty$,

$$u(\cdot,t) \to v(\cdot) \qquad \text{strongly in } W^{1,p}(0,1), \tag{3.18}$$

$$u_t(\cdot,t) \to 0 \qquad \text{strongly in } W^{1,2}(0,1), \tag{3.19}$$

for all $p > 1$, where $\{v, 0\}$ is a rest point of (3.11)-(3.13). Thus solutions to the dynamical equations do not mimic the typical behaviour of minimizing sequences. The results of Pego do not seem, however, to be sufficient to establish whether or not a version of the prototheorem holds.

Example 3.3. *(The Becker-Döring cluster equations)*
These are the infinite set of ordinary differential equations

$$\begin{aligned} \dot{c}_r &= J_{r-1}(c(t)) - J_r(c(t)), \qquad r \geq 2, \\ \dot{c}_1 &= -J_1(c(t)) - \sum_{r=1}^{\infty} J_r(c(t)), \end{aligned} \tag{3.20}$$

where $c(t)$ denotes the infinite vector $(c_r(t))$,

$$J_r(c) = a_r c_1 c_r - b_{r+1} c_{r+1}, \tag{3.21}$$

and the coefficients $a_r > 0$, $b_r > 0$ are constant. The physical significance of (3.20) is discussed in the article in this volume by Carr [15].

Let $X = \{y = (y_r) : \| y \| \overset{\text{def}}{=} \sum_{r=1}^{\infty} r \mid y_r \mid < \infty\}$. X is a Banach space with the indicated norm. Solutions of (3.20) are sought as continuous functions $c : [0, \infty) \to X^+$, where

$$X^+ = \{y \in X : y_r \geq 0, \quad r = 1, 2, \ldots\}. \tag{3.22}$$

The system (3.20) possesses the Lyapunov function

$$V(c) = \sum_{r=1}^{\infty} c_r \left(\ln \left(\frac{c_r}{Q_r} \right) - 1 \right), \tag{3.23}$$

where $Q_1 = 1$, $Q_{r+1}/Q_r = a_r/b_{r+1}$, and there is a constant of motion, the density

$$\rho = \sum_{r=1}^{\infty} rc_r. \tag{3.24}$$

For suitable coefficients a_r, b_r there exists $\rho_s > 0$ such that there is a unique rest point $c^{(\rho)}$ of (3.20) with density ρ for $\rho \in [0, \rho_s]$, and no rest point with any density $\rho > \rho_s$. Furthermore $c^{(\rho)}$ is the unique absolute minimizer of the problem

$$\underset{c \in X^+,\ \sum_{r=1}^{\infty} rc_r = \rho}{\text{Minimize}} \quad V(c)\,. \tag{3.25}$$

The equations (3.20) were analyzed in Ball, Carr & Penrose [8], Ball & Carr [7]; see also Ball [2] for remarks on the variational problem (3.25). It follows from [8],[7] that under suitable hypotheses on the a_r, b_r the conclusion of the prototheorem holds. That is, given $c(0) \in X^+$ with $\sum_{r=1}^{\infty} rc_r(0) = \rho$, and any sequence $t_j \to \infty$, $c(t_j)$ is a minimizing sequence for (3.25). Note that this conclusion holds even in the case $\rho > \rho_s$, when the minimum in (3.25) is not attained.

Example 3.4. (*Model equations related to phase transitions in solids*)
In Example 3.2, the Lyapunov function V given by (3.15) has minimizing sequences that oscillate more and more finely, converging weakly to a state that is not a minimizer. On the other hand there are minimizing sequences which do not behave like this, consisting, for example, of a single minimizer. The results of Pego show that the dynamics chooses to imitate the latter kind of minimizing sequence rather than the former. In the crystal problem described in Example 2.2 minimizing sequences are forced to oscillate more and more finely, leading to interesting possibilities for a corresponding dynamical model. Does the dynamics imitate the minimizing sequences, or is it still the case that all solutions tend to equilibria? This is a formidable problem, so it makes sense to first try out some one-dimensional examples. The most obvious candidate is the problem

$$u_{tt} = (\sigma(u_x) + u_{xt})_x - 2u, \qquad 0 < x < 1, \tag{3.26}$$

with boundary conditions

$$u = 0 \text{ at } x = 0, 1, \tag{3.27}$$

and initial conditions

$$u(x,0) = u_0(x), \qquad u_t(x,0) = u_1(x), \qquad 0 < x < 1. \tag{3.28}$$

As before, assume that

$$\sigma(u_x) = W'(u_x), \qquad W(u_x) = (u_x^2 - 1)^2. \tag{3.29}$$

Then $V(u, u_t)$ is a Lyapunov function for (3.26)-(3.28), where

$$V(u,p) = \int_0^1 [\frac{1}{2}p^2 + W(u_x) + u^2]\,dx. \qquad (3.30)$$

The minimizing sequences of V subject to (3.27) all oscillate faster and faster, converging weakly but not strongly to $\{0,0\}$ in $W_0^{1,4}(0,1) \times L^2(0,1)$. (See the paper in this volume by Müller [27] for a study of this variational problem with surface energy added.)

The problem (3.26)-(3.28) has been studied in joint work of P.J.Holmes, R.D.James, R.L.Pego, P.Swart and the author [9], together with the much more tractable problem consisting of the equation

$$u_{tt} = \left(\int_0^1 u_x^2\,dx - 1\right) u_{xx} + u_{xxt} - 2u, \qquad 0 < x < 1, \qquad (3.31)$$

with boundary and initial conditions (3.27),(3.28). This problem has the Lyapunov function $\overline{V}(u, u_t)$, where

$$\overline{V}(u,p) = \int_0^1 [\frac{1}{2}(p^2 - u_x^2) + u^2]\,dx + \frac{1}{4}\left(\int_0^1 u_x^2\,dx\right)^2. \qquad (3.32)$$

There are countably many rest points of (3.31),(3.27) given by

$$u_k = a_k \sin k\pi x, \qquad k \text{ an integer}, \qquad (3.33)$$

for suitable coefficients a_k. It can easily be proved that

$$\inf_X \overline{V} = -\frac{1}{4}, \qquad (3.34)$$

where $X = H_0^1(0,1) \times L^2(0,1)$. Then we have the result

Theorem 2 *Let u be any weak solution of* (3.31),(3.27). *As $t \to \infty$ either*
(i) $\{u, u_t\} \to \{u_k, 0\}$ *strongly in X for some k, or*
(ii) $\{u, u_t\} \rightharpoonup \{0, 0\}$ *weakly in X, but not strongly, and*

$$\lim_{t\to\infty} \overline{V}(t) = -\frac{1}{4}. \qquad (3.35)$$

The alternatives (i),(ii) *both occur for dense sets of initial data in X, the set corresponding to* (ii) *being of second category.*

By contrast, for the problem (3.26)-(3.28) it is shown in [9] that there is no solution $\{u, u_t\}$ for which

$$\lim_{t\to\infty} V(t) = 0, \qquad (3.36)$$

i.e. no solution which realizes an absolute minimizing sequence.

References

[1] G. Andrews and J. M. Ball. Asymptotic behaviour and changes of phase in one-dimensional nonlinear viscoelasticity. *J. Differential Eqns*, 44:306–341, 1982.

[2] J. M. Ball. Loss of the constraint in convex variational problems. In *Analyse Mathématique et Applications; Contributions en l'Honneur de J.-L.Lions*, pages 39–53, Gauthier-Villars, 1988.

[3] J. M. Ball. Material instabilities and the calculus of variations. In M. E. Gurtin, editor, *Phase transformations and material instabilities in solids*, pages 1–20, Mathematics Research Center, University of Wisconsin, Academic Press, 1984.

[4] J. M. Ball. Minimizing sequences in thermomechanics. In *Proc. Meeting on 'Finite Thermoelasticity'*, pages 45–54, Accademia Nazionale dei Lincei, Roma, 1986.

[5] J. M. Ball. On the asymptotic behaviour of generalized processes, with applications to nonlinear evolution equations. *J. Differential Eqns*, 27:224–265, 1978.

[6] J. M. Ball. Saddle-point analysis for an ordinary differential equation in a Banach space, and an application to dynamic buckling of a beam. In R. W. Dickey, editor, *Nonlinear Elasticity*, pages 93–160, Mathematics Research Center, University of Wisconsin, Academic Press, 1973.

[7] J. M. Ball and J. Carr. Asymptotic behaviour of solutions of the Becker-Döring equations for arbitrary initial data. *Proc. Royal Soc. Edinburgh A*, 108:109–116, 1988.

[8] J. M. Ball, J. Carr, and O. Penrose. The Becker-Döring cluster equations; basic properties and asymptotic behaviour of solutions. *Comm. Math. Phys.*, 104:657–692, 1986.

[9] J. M. Ball, P. J. Holmes, R. D. James, R. L. Pego, and P. Swart. to appear.

[10] J. M. Ball and R. D. James. Fine phase mixtures as minimizers of energy. *Arch. Rat. Mech. Anal.*, 100:13–52, 1987.

[11] J. M. Ball and R. D. James. Proposed experimental tests of a theory of fine microstructure, and the two-well problem. to appear.

[12] J. M. Ball and G. Knowles. Lyapunov functions for thermoelasticity with spatially varying boundary temperatures. *Arch. Rat. Mech. Anal.*, 92:193–204, 1986.

[13] J. M. Ball and M. Slemrod. Nonharmonic Fourier series and the stabilization of distributed bilinear control systems. *Comm. Pure Appl. Math.*, 32:555–587, 1979.

[14] E. A. Barbashin and N. N. Krasovskii. Stability of motion in the large. *Dokl. Akad. Nauk SSSR*, 86:453–456, 1952.

[15] J. Carr. Dynamics of cluster growth. In *this proceedings*.

[16] J. Carr and R. L. Pego. Metastable patterns in solutions of $u_t = \epsilon^2 u_{xx} - f(u)$. *Comm. Pure Appl. Math.*, to appear.

[17] M. Chipot and D. Kinderlehrer. Equilibrium configurations of crystals. *Arch. Rat. Mech. Anal.*, 103:237–277, 1988.

[18] B. D. Coleman and E. H. Dill. On thermodynamics and the stability of motion of materials with memory. *Arch. Rat. Mech. Anal.*, 51:1–53, 1973.

[19] C. M. Dafermos. Asymptotic behaviour of solutions of evolution equations. In M. G. Crandall, editor, *Nonlinear Evolution Equations*, pages 103–124, Mathematics Research Center, University of Wisconsin, Academic Press, 1978.

[20] P. Duhem. *Traité d'Énergetique ou de Thermodynamique Générale.* Gauthier-Villars, Paris, 1911.

[21] J. L. Ericksen. Thermoelastic stability. In *Proc* 5th *National Cong. Appl. Mech.*, pages 187–193, 1966.

[22] I. Fonseca. Interfacial energy and the Maxwell rule. *Arch. Rat. Mech. Anal.*, 106:63–95, 1989.

[23] J. K. Hale. Dynamical systems and stability. *J. Math. Anal. Appl.*, 26:39–59, 1969.

[24] D. Henry. *Geometric Theory of Semilinear Parabolic Equations.* Volume 840 of *Lecture Notes in Mathematics*, Springer-Verlag, 1981.

[25] J. P. LaSalle. The extent of asymptotic stability. *Proc. Nat. Acad. Sci. USA*, 46:363–365, 1960.

[26] P. Lin. Maximization of the entropy for an elastic body free of surface traction. *to appear.*

[27] S. Müller. Minimizing sequences for nonconvex functionals, phase transitions and singular perturbations. In *this proceedings*.

[28] S. Müller. Strong convergence and arbitrarily slow decay of energy for a class of bilinear control problems. *J. Differential Eqns*, to appear.

[29] G. P. Parry. On shear bands in crystals. *J. Mech. Phys. Solids*, 35:367–382, 1987.

[30] R. L. Pego. Phase transitions in one-dimensional nonlinear viscoelasticity: admissibility and stability. *Arch. Rat. Mech. Anal.*, 97:353–394, 1987.

DISCONTINUOUS SOLUTIONS OF BOUNDED VARIATIONS TO PROBLEMS OF THE CALCULUS OF VARIATIONS AND OF QUASI LINEAR HYPERBOLIC DIFFERENTIAL EQUATIONS. INTEGRALS OF SERRIN AND WEIERSTRASS.

Lamberto Cesari
Department of Mathematics, University of Michigan
Ann Arbor, Michigan 48109

I. BV FUNCTIONS OF $\nu \geq 1$ INDEPENDENT VARIABLES.

In 1936 Cesari [5] introduced a concept of real functions $z : G \to \mathbf{R}$, or $z(t)$, or $z(t^1, \ldots, t^\nu)$, of bounded variation (BV) from a domain G of $\mathbf{R}^\nu$. For the case $\nu = 2$, G the rectangle $(a, b; c, d)$, the definition is very simple: we say that z is BV in $G = (a, b; c, d)$ provided $z \in L_1(G)$ and there is a set E of measure zero in G such that the total variation $V_x(y)$ of $z(\cdot, y)$ in (a, b) is of class $L_1(c, d)$, and the total variation $V_y(x)$ of $z(x, \cdot)$ in (c, d) is of class $L_1(a, b)$, where these total variations are computed completely disregarding the values taken by z in E. The number

$$V_0 = V_0(z, G) = \int_a^b V_y(x)\, dx + \int_c^d V_x(y)\, dy$$

may well be taken as a definition of total variation of z in $G = (a, b; c, d)$, (with respect to such a set $E \subset G$ of measure zero). Analogous definitions hold for BV functions $z(t^1, \ldots, t^\nu)$ in an interval G of $\mathbf{R}^\nu$.

We omit here the more involved definition of BV functions in a general bounded domain G of $\mathbf{R}^\nu$.

If z is continuous in G, then no set E need be considered and the concept reduces to Tonelli's concept of BV continuous functions. For discontinuous functions, examples show how essential it is to disregard sets E of measure zero in G. On the other hand, the concept obviously concerns equivalent classes in $L_1(G)$.

We may think of $z(t)$, $t \in G \subset \mathbf{R}^\nu$, as defining a nonparametric possibly discontinuous surface, $S : z = z(t)$, $t \in G$, in $\mathbf{R}^{\nu+1}$, and we may take as generalized Lebesgue area $L(S)$ of S the lower limit of the elementary areas $a(\sum)$ of the polyhedral surfaces $\sum : z = Z(t)$, $t \in G$, converging to z pointwise a.e. in G (or in $L_1(G)$). More precisely, if $(\sum_k)$ denotes any sequence of polyhedral surfaces $\sum_k : z = z_k(t)$, $T \in G$ converging to z pointwise a.e. in G (or in $L_1(G)$),we take for $L(S)$ the number, $0 \leq L(S) \leq +\infty$, defined by

$$L(S) = \operatorname*{Inf}_{(\sum_k)} \varliminf_{k \to \infty} a(\textstyle\sum_k).$$

Cesari proved [5] that $L(S)$ is finite if and only if z is BV in G. This shows that the concept of BV functions is independent of the direction of the axes in $\mathbb{R}^\nu$. More than that, the concept of BV functions is actually invariant with respect to $1-1$ continuous transformations in $\mathbb{R}^\nu$ which are Lipschitzian in both directions.

In 1937 Cesari [6a] proved that for $\nu = 2$, $G = (0, 2\pi; 0, 2\pi)$ and z BV in G, then the double Fourier series of z converges to z (by rectangles, by lines and by columns) a.e. in G. Comparable, though weaker, results hold for BV functions of $\nu > 2$ independent variables and their multiple Fourier series [6b].

In 1950 Cafiero [4] and later in 1957 Fleming [15] proved the relevant compactness theorem: any sequence (z_k) of BV functions with equibounded total variations, say $V_0(z_k, G) \leq C$, and equibounded mean values in G, possesses a subsequence z_{k_s} which is pointwise convergent a.e. in G as well as strongly convergent in $L_1(G)$ toward a BV function z.

In 1967 Conway and Smoller [12] used these BV functions in connection with the weak solutions (shock waves) of conservation laws, a class of nonlinear hyperbolic partial differential equations in $\mathbb{R}^+ \times \mathbb{R}^\nu$. Indeed they proved that, if the Cauchy data on $(0) \times \mathbb{R}^\nu$ are locally BV, then there is a unique weak solution on $\mathbb{R}^+ \times \mathbb{R}^\nu$, also locally BV and satisfying an entropy condition. Without any entropy condition there are in general infinitely many weak solutions. Analogous results for $\nu = 1$ had been obtained before by Oleinik [18]. Later, Dafermos [13] and Di Perna [14] characterized the properties of the BV weak solutions of conservation laws.

Meanwhile, in the fifties, distribution theory became known, and in 1957 Krickeberg [17] proved that the BV functions are exactly those $L_1(G)$ functions whose first order partial derivatives in the sense of distributions are finite measures in G.

Thus a BV function $z(t)$, $t \in G$, G a bounded domain in $\mathbb{R}^\nu$, possesses first order partial derivatives in the sense of distributions which are finite measures μ_j, $j = 1, \dots, \nu$. On the other hand, if we think of the initial definition of z, we see that the set E of measure zero in G has intersection $E \cap \ell$ of linear measure zero on almost all lines ℓ parallel to the axes. Hence z is BV on almost all such straight lines when we disregard the values taken by z on E, and therefore has "usual" partial derivatives $D^j z$ a.e. in G, and these derivatives are functions in G of class $L_1(G)$. We call these $D^j z(t)$, $t \in G$, $j = 1, \dots, \nu$, computed by usual incremental quotients disregarding the values taken by z on E, the generalized first order partial derivatives of z in G.

Much work followed on BV functions in terms of the new definition, that is, thought of as those $L_1(G)$ functions whose first order derivatives are finite measures. We mention here Fleming [15], Volpert [22], Gagliardo [16], Anzellotti and Giaquinta ([1]), and also De Giorgi, Da Prato, Ferro, Caligaris, Oliva, Fusco, Temam. However, there are advantages in using both view points.

Great many properties of BV functions have been proved. To begin with, a "total variation" $V(z, G)$ can be defined globally in terms of functional analysis,

$$V(z, G) = \operatorname{Sup}\left[\left(\int_G f_1 d\mu_1\right)^2 + \ldots + \left(\int_G f_\nu d\mu_\nu\right)^2\right]^{1/2},$$

where the Sup is taken for all $f_1, \dots, f_\nu \in C(G)$ with $f_1^2 + \ldots + f_\nu^2 \leq 1$ and compact support in G.

If (z_k) is a sequence of BV functions on G with equibounded total variations, say $V(z_k, G) \leq C$, and $z_k \to z$ in $L_1(G)$, then z is BV and $V(z,G) \leq \underline{\lim}_{k\to\infty} V(z_k, G)$.

The question of the existence of traces $\gamma : \partial G \to \mathbf{R}$ for BV functions $z : G \to \mathbf{R}$ has been discussed under both view points. Note that for a BV function z in an interval $G = (a, b; c, d)$ it is trivial that the generalized limits $z(a+, y)$ and $z(b-, y)$, $z(x; c+)$ and $z(x, d-)$, exist a.e. and are L_1 functions, i.e., the trace $\gamma(z)$ of z on ∂G exists and is $L_1(\partial G)$. For general domains G in $\mathbf{R}^\nu$ possessing the cone property everywhere on ∂G, a theorem of Gagliardo [16] characterizes the properties of ∂G, and one can prove that any BV function z in a bounded domain G with the cone property and $\mathcal{H}^{\nu-1}(\partial G) < \infty$ possesses a trace $\gamma(z)$ on ∂G with $\gamma(z) \in L_1(\partial G)$.

We mention here the following theorem by Gagliardo on bounded domains G with the cone property: If G is a bounded open domain in $\mathbf{R}^\nu$ having the cone property, then there is a finite system $(G_1, \ldots, G_m)$ of open subsets of G with max diam G_s as small as we want, each G_s has the cone property and has locally Lipschitzian boundary ∂G_s.

From this result, and trace properties for Lipschitzian domains, it is possible to define the trace $\gamma(z)$ of a BV function on ∂G, for G bounded and with the cone property. An equivalent definition of traces of BV functions in terms of the distributional definition is also well known.

We come now to the delicate question of the continuity of the traces of $\gamma(z)$ of BV functions z in a domain G, in other words whether $z_k \to z$, say in $L_1(G)$, may actually imply—under assumptions—that $\gamma(z_k) \to \gamma(z)$ in $L_1(\partial G)$. A number of devices have been proposed to this effect. For instance, Anzellotti and Giaquinta ([1]) have recently proved the following statement in terms of the distributional definition of BV functions: If G has the cone property at every point of ∂G, if $\mathcal{H}^{\nu-1}(\partial G) < \infty$, if the functions z_k are BV with $V(z_k) \leq C$, if $z_k \to z$ in $L_1(G)$ with $V(z_k) \to V(z)$, then $\gamma(z_k) \to \gamma(z)$ in $L_1(\partial G)$. A parallel proof of this statement is available in terms of the original definition of BV functions. We mention here that it is well known that any BV function $z(t)$, $t \in G \subset \mathbf{R}^\nu$, can be approximated in $L_1(G)$ by BV smooth functions z_k with $V(z_k) \to V(z)$, hence $V(z_k) \leq C$.

We shall see now how these ideas have been used in questions of optimization.

II. CALCULUS OF VARIATIONS IN CLASSES OF BV FUNCTIONS.

When the state variable z, or $z(t) = (z^1, \ldots, z^m)$, $t \in G \subset \mathbf{R}^\nu$, is only BV, that is, each component z^i is BV in G, the usual Lebesgue integral of the calculus of variations

$$I(z) = \int_G f_0(t, z(t), Dz(t))\, dt,$$

$$t = (t^1, \ldots, t^\nu) \in G \subset \mathbf{R}^\nu,\ z(t) = (z^1, \ldots, z^m),\ \nu \geq 1,\ m \geq 1,$$

may not give a true, or stable value for the functional of interest. There are two basic processes to determine a true, or stable value for the underlying functional, and both have generated a great deal of recent work.

One is the limit process already proposed by *Weierstrass*, leading to a functional, or Weierstrass integral, W(z). Tonelli made use of it in his early work (1914) on the direct method in the calculus of variations on parametric continuous curves C, or $z(t) = (z^1, \ldots, z^m)$, $a \leq t \leq b$, of finite length, hence all z^i are BV and continuous. Recently, Cesari [8ab] presented an abstract formulation of the Weierstrass integral as a Burkill-type limit on "quasi additive" set functions $\phi(I) = (\phi_1, \ldots, \phi_n)$, $I \subset G$, and state functions $z(t) = (z^1, \ldots, z^m)$, $t \in G \subset \mathbf{R}^\nu$. Cesari also proved [8b] that $W(z)$ has a representation as a Lebesgue-Stieltjes integral in terms of measures and Radon-Nikodym derivatives derived from the set function ϕ. Warner [23] then proved lower semicontinuity theorems for continuous varieties, and very recently Brandi and Salvadori [1defg] extended further the abstract formulation, proved further representation properties, and lower semicontinuity theorems, both in the parametric and in the non-parametric case, and for vector functions z (state variables) only BV, possibly discontinuous, possibly not Sobolev (see Section VII below).

Another approach was proposed by *Serrin* [20a] leading to a functional, or Serrin integral, $\mathcal{T}(z)$, in classes of BV vector functions $z(t)$, $t \in G \subset \mathbf{R}^\nu$. The Serrin functional $\mathcal{T}(z)$ is obtained by taking lower limits on the values of I on AC, or $W^{1,1}(G)$ functions, a process which is similar to the one with which Lebesgue area is defined. Recently, Cesari, Brandi and Salvadori [10ab] proved closure and lower closure theorems, hence theorems of lower semicontinuity in the L_1-topology, and finally theorems of existence of the absolute minimum of $\mathcal{T}(z)$ in classes of BV vector functions whose total variations $V(z)$ are equibounded [10ab]. We proved also that $I(z) \leq \mathcal{T}(z)$, and that $\mathcal{T}$ is a proper extension of I in the sense that $\mathcal{T}(z) = I(z)$ for all z which are AC, or $W^{1,1}(G)$ (see Sections III,IV below). A number of applications of this approach has been announced [9abc, 11ab].

III. PROBLEMS OF OPTIMIZATION FOR SIMPLE INTEGRALS, $\nu = 1$, BY THE USE OF SERRIN'S FUNCTIONAL.

We may be interested either in problems of the classical calculus of variations involving a vector valued state variable $z(t) = (z^1, \ldots, z^n)$, $t_1 \leq t \leq t_2$, or in problems of optimal control involving an analogous state variable $z(t) = (z^1, \ldots, z^n)$ and a control variable $u(t) = (u^1, \ldots, u^m)$, $t_1 \leq t \leq t_2$, with given control space $U(t, z)$ and constraint $u(t) \in U(t, z(t))$.

It is more general and more satisfactory (cfr. [7b]), to deparametrize the problems of optimal control, and concern ourselves exclusively with generalized problems of the calculus of variations with constraints on the derivatives, say

$$
\begin{aligned}
&I(z) = \int_{t_1}^{t_2} f_0(t, z(t), z'(t))\, dt = \text{minimum}, \\
&(t, z(t)) \in A \subset \mathbf{R}^{n+1},\ z'(t) \in Q(t, z(t)), \qquad (1)
\end{aligned}
$$

where $t \in [t_1, t_2] \subset \mathbf{R}$ (a.e.), where A is a subset of $\mathbf{R}^{n+1}$ whose projection on the t-axis contains $[t_1, t_2]$, and where, for every $(t, z) \in A$, a set $Q(t, z)$ is given constraining the direction $z'(t)$ of the tangent to the state variable z a.e. in $[t_1, t_2]$. The process of deparametrization has been discussed in detail in [7b].

For what concerns boundary conditions for problem (1), we restrict ourselves here to Dirichlet type boundary conditions

$$z(t_1) = z_1, \; z(t_2) = z_2. \tag{2}$$

Above, let M denote the set $M = [(t, z, \xi)] | (t, z) \in A, \xi \in Q(t, z)] \subset \mathbf{R}^{1+2n}$, and let $f_0(t, z, \xi)$ be a real valued function on M. Let Ω be a class of admissible functions, i.e., functions $z : [t_1, t_2] \to \mathbf{R}^n$, or $z(t) = (z^1, \ldots, z^n)$, such that (i) z is BV in $[t_1, t_2]$; (ii) $(t, z(t)) \in A$, $z'(t) \in Q(t, z(t))$ a.e. in $[t_1, t_2]$; (iii) $f_0(\cdot, z(\cdot), z'(\cdot)) \in L_1[t_1, t_2]$.

It is easy to see that the Lebesgue integral definition (1) of the functional (I) does not yield stable and realistic values for I, and one may use a Serrin type integral. To this effect, for every $z \in \Omega$ we denote by $\Gamma(z)$ the class of all sequences (z_k) of elements $z_k \in \Omega$ with (a) z_k is AC in $[t_1, t_2]$; (b) $z_k \to z$ pointwise a.e. in $[t_1, t_2]$.

If $\Gamma(z)$ is empty we take $\mathcal{T}(z) = +\infty$. If $\Gamma(z)$ is not empty, then we take

$$\mathcal{T}(z) = \inf \underline{\lim} \int_{t_1}^{t_2} f_0(t, z_k(t), z_k'(t))\, dt = \inf_{\Gamma(z)} \underline{\lim}_{k \to \infty} I(z_k). \tag{3}$$

This is a Serrin type definition of the functional which was inspired to the Lebesgue area.

If problem (1) has assigned boundary conditions, say of the Dirichlet type 2, then let $\Gamma(z)$ denote the class of all sequences (z_k) of elements z_k in Ω with (a) z_k is AC and satisfies the boundary conditions; (b$'$) $z_k \to z$ pointwise a.e. in $[t_1, t_2]$, in particular $z_k(t_i) \to z(t_i)$, $i = 1, 2$. Then the analogous integral defined by (3) could be denoted by $\mathcal{T}^*$ and obviously $\mathcal{T} \leq \mathcal{T}^*$.

We can now state a lower semicontinuity theorem and an existence theorem for the integrals I and $\mathcal{T}$ on BV functions. To this purpose we have first to define as usual the "augmented" sets $\tilde{Q}(t, z)$ as follows:

$$\tilde{Q}(t, z) = [(\tau, \xi) | \tau \geq f_0(t, z, \xi), \; \xi \in Q(t, z)] \subset \mathbf{R}^{n+1}.$$

A lower semicontinuity theorem. Let us assume that (i) A is closed; (ii) the sets $\tilde{Q}(t, z)$ are closed, convex, and satisfy property (Q) with respect to (t, z) at every $(t, z) \in A$; (iii) $f_0(t, z, \xi)$ is lower semicontinuous in M, and there exists some function $\lambda \in L_1[t_1, t_2]$ such that $f_0(t, z, \xi) \geq \lambda(t)$ for all $(t, z, \xi) \in M$. Let $z(t)$, $t \in [t_1, t_2]$, be BV, and let $z_k(t)$, $t \in [t_1, t_2]$, $k = 1, 2, \ldots$, be a sequence of AC functions z_k such that $z_k \to z$ pointwise a.e. in $[t_1, t_2]$, $(t, z_k(t)) \in A$, $z_k'(t) \in Q(t, z_k(t))$ a.e. in $[t_1, t_2]$, and $V(z_k) \leq C$. Then, $(t, z(t)) \in A, z'(t) \in Q(t, z(t))$ a.e. in $[t_1, t_2]$, and $I(z) \leq \underline{\lim}_{k \to \infty} I(z_k)$ [10b].

A fundamental consequence of this lower semicontinuity theorem is that if z_k is any of the sequences of AC elements in $\Gamma(z)$, with $V(z_k) \leq C$, and we take $j = \underline{\lim}_{k \to \infty} I(z_k)$, then

$$I(z) \leq \mathcal{T}(z) \leq j = \underline{\lim}_{k \to \infty} I(z_k).$$

Furthermore, the Serrin integral $\mathcal{T}$ is actually an extension of the integral I. Indeed, if $z \in \Omega \cap AC$, then, by taking $z_k = z$ we conclude that $I(z) \le \mathcal{T}(z) \le \underline{\lim}_{k\to\infty} I(z_k) = I(z)$.

Note that for sequences (z_k) as above with $V(z_k)$ unbounded, it may well occur that $\mathcal{T}(z) < I(z)$ as it has been proved by examples (cfr. [10b]).

We mention here that Kuratowski's property (K) at a point (t_0, z_0) is expressed by the relation

$$\tilde{Q}(t_0, z_0) = \cap_{\delta>0}\, cl\, [\cup \tilde{Q}(t, z),\ (t - t_0)^2 + |z - z_0|^2 \le \delta^2].$$

The analogous condition (Q) at the point (t_0, z_0) is expressed by the relation

$$\tilde{Q}(t_0, z_0) = \cap_{\delta>0}\, cl\, co\, [\cup \tilde{Q}(t, z),\ (t - t_0)^2 + |z - z_0|^2 \le \delta^2].$$

If problem (1) has assigned boundary conditions of the type (2), then in the theorem above we assume that $z_k \to z$ a.e. in $[t_1, t_2]$, in particular that $z_k(t_i) = z(t_i)$, $i = 1, 2$, and the same statement holds for $\mathcal{T}^*$.

An existence theorem for the integral $\mathcal{T}$. Let us assume that (i) A is compact and M is closed; (ii) the sets $\tilde{Q}(t, z)$ are closed, convex, and satisfy property (Q) with respect to (t, z) at every point (t, z) of A; (iii) $f_0(t, z, \xi)$ is lower semicontinuous in M. Assume that the class Ω is nonempty and closed, $V(z) \le C$ for all $z \in \Omega$, and $\Gamma(z)$ is nonempty for at least one z. Then the functional $\mathcal{T}$ has an absolute minimum $z \in BV$ in Ω [10b].

In other words, let i denote the infimum of $I(z)$ for $z \in AC \cap \Omega$, let (z_k) denote a sequence of elements $z_k \in AC \cap \Omega$ with $I(z_k) \to i$. Then, there is an element $z \in \Omega$, $z \in BV$, such that $I(z) \le \mathcal{T}(z) = i$.

Examples have been given in [10a].

IV. PROBLEMS OF OPTIMIZATION FOR MULTIPLE INTEGRALS AND BV DISCONTINUOUS FUNCTIONS, $\nu > 1$, BY THE USE OF SERRIN'S FUNCTIONAL.

Let $\nu > 1$, $n \ge 1$, and let $G \subset \mathbf{R}^\nu$ be a bounded domain in the t-space $\mathbf{R}^\nu$, $t = (t^1, \ldots, t^\nu)$, possessing the cone property at every point of its boundary ∂G. Let $A \subset \mathbf{R}^{\nu+n}$ be a compact subset of the tz-space $\mathbf{R}^{\nu+n}$, whose projection on the t-space contains G.

We shall deal with vector valued functions $z(t) = (z^1, \ldots, z^n)$, z^i BV in G, therefore possessing first order partial derivatives in the sense of distributions which are measures $\mu_{ij}, j = 1, \ldots, \nu$, $i = 1, \ldots, n$, and in addition also generalized first order derivatives $D^j z^i$ a.e. in G, as functions of class $L_1(G)$, which are obtained as limits of incremental quotients when we disregard the values taken by the functions in suitable sets E of measure zero in G. We may need only a subset of such derivatives $D^j z^i$ as follows.

For every $i = 1, \ldots, n$, let $\{j\}_i$ be a system of indices $1 \le j_1 < \ldots < j_s \le \nu$, let $D^j z^i$, $j \in \{j\}_i$, denote the corresponding system of first order partial derivatives of the function z^i, and let N be their total number. Then by Dz we denote the N-vector function $Dz(t) = (D^j z^i, j \in \{j\}_i, i = 1, \ldots, n)$, $t \in G$ (a.e.).

For every $(t, z) \in A$ let $Q(t, z)$ be a given subset of $\mathbf{R}^N$. Let $M \subset \mathbf{R}^{\nu+n+N}$ denote the set $M = [(t, z, \xi)|(t, z) \in A, \; \xi \in Q(t, z)]$, and let $f_0(t, z, \xi)$ be a given real-valued function in M. We are interested in the multiple integral problem of the calculus of variations with constraints on the derivatives

$$I(z) = \int_G f_0(t, z(t), Dz(t))\, dt = \text{minimum},$$
$$(t, z(t)) \in A, \; Dz(t) \in Q(t, z(t)), \; t \in G \; (a.e.), \tag{1}$$

and possible Dirichlet type boundary conditions of the form $z(t) = \phi(t), t \in \partial G(\mathcal{H}^{\nu-1} - a.e.)$ on ∂G. Again we introduce a Serrin type integral.

Let Ω be a class of admissible functions $z(t) = (z^1, \ldots, z^n)$, $t \in G$, such that (i) z is BV in G; (ii) $(t, z(t)) \in A, Dz(t) \in Q(t, z(t)), t \in G(a.e.)$; (iii) $f_0(\cdot, z(\cdot), Dz(\cdot)) \in L_1(G)$.

To simplify notations, let AC, or $AC(G)$, denote the class of functions $z(t) = (z^1, \ldots, z^n)$, $t \in G$, whose components z^i are of Sobolev class $W^{1,1}(G)$, or briefly, Beppo Levi functions.

For any element $z \in \Omega$ let $\Gamma(z)$ denote the class of all sequences (z_k) of elements z_k in Ω with (a) z_k is AC in G; (b) $z_k \to z$ strongly in $L_1(G)$.

If $\Gamma(z)$ is empty we take $\mathcal{T}(z) = +\infty$. If $\Gamma(z)$ is not empty we take

$$\mathcal{T}(z) = \inf \underline{\lim} \int_G f_0(t, z_k(t), Dz_k(t))\, dt = \inf_{\Gamma(z)} \underline{\lim}_{k \to \infty} I(z_k). \tag{2}$$

To state an existence theorem we introduce, as usual, the augmented sets $\tilde{Q}(t, z) \subset \mathbf{R}^{N+1}$ as follows

$$\tilde{Q}(t, z) = [(\tau, \xi)|\tau \ge f_0(t, z, \xi), \; \xi \in Q(t, z)].$$

Beside property (Q) we shall require on the sets $\tilde{Q}(t, z)$ another property, or property $\bar{F}_1$ [10c].

We say that the sets $Q(t, z)$, $(t, z) \in A$, have property $\bar{F}_1$ with respect to z at the point $(t_0, z_0) \in A$ provided, given any number $\sigma > 0$, there are constants $C > 0$, $\delta > 0$ which depend on t_0, z_0, σ, such that for any set of measurable vector functions $\eta(t), z(t), \xi(t), t \in H$, on a measurable subset H of points t of G with $(t, z(t)) \in A$, $|z(t) - z_0| > \sigma$, $(\eta(t), \xi(t)) \in \tilde{Q}(t, z(t))$ for $t \in H$, $|t - t_0| \le \delta$, there are other measurable vector functions $\bar{\eta}(t)$, $\bar{z}(t)$, $\bar{\xi}(t)$, $t \in H$, such that

$$(t, \bar{z}(t)) \in A, \; |\bar{z}(t) - z_0| \le \sigma, \; (\bar{\eta}(t), \bar{\xi}(t)) \in \tilde{Q}(t, \bar{z}(t)),$$
$$|\xi(t) - \bar{\xi}(t)| \le C[|z(t) - \bar{z}(t)| + |t - t_0|],$$
$$\bar{\eta}(t) \le \eta(t) + C[|z(t) - \bar{z}(t)| + |t - t_0|] \text{ for } t \in H$$

We denote by $\bar{F}_2$ the same condition with $\bar{z}(t) = z_0$. These conditions are inspired to analogous ones proposed by Rothe, Berkovitz, Browder (cfr. Cesari [7b], Section 13). These conditions have been replaced by more general ones in [10d].

An existence theorem. Let us assume that (i) A is compact and M is closed; (ii) the sets $\tilde{Q}(t,z)$ are closed, convex and satisfy properties (Q) and $(\tilde{F}_1)$ at every point $(t,z) \in A$; (iii) $f_0(t,z,\xi)$ is bounded below and lower semicontinuous in (t,z,ξ). Also, let us assume that the class Ω is nonempty and closed, and $\Gamma(z)$ is nonempty for at least one $z \in \Omega$. Then the functional $\mathcal{T}$ has an absolute minimum z in Ω, $z \in BV$ in G [10c].

In other words, let i denote the infimum of $I(z)$ for $z \in AC \cap \Omega$, let (z_k) denote any sequence of elements $z_k \in AC \cap \Omega$ with $I(z_k) \to i$. Then there is at least one element $z \in \Omega$, $z \in BV$, such that $I(z) \le \mathcal{T}(z) = i$.

V. EXISTENCE OF BV POSSIBLY DISCONTINUOUS ABSOLUTE MINIMA FOR CERTAIN INTEGRALS WITHOUT GROWTH PROPERTIES.

Recently I considered [9a] multiple integrals of the form

$$I(z) = \int_G \sum_{i=1}^{m} \Big| \sum_{j=1}^{\nu} [U_{ij}(t,z)]_{t_j} + V_i(t,z) \Big| \, dt,$$

$$z(t) = (z_1, \ldots, z_m),\ t = (t_1, \ldots, t_\nu) \in G \subset \mathbf{R}^\nu,$$

$$z(t) = w(t),\ t \in B \subset \partial G, \tag{1}$$

and associated Serrin functionals $\mathcal{T}(z)$. I studied these integrals in classes of BV vector functions $z(t) = (z_1, \ldots, z_m)$, $t \in G$, with equibounded total variations. Here the U_{ij} are given functions of class C^1 and the V_i are given locally Lipschitzian functions. The existence theorems we mentioned in Section IV above, and we had proved in [10c], do not apply directly to the integrals (1). However, I proved in [9a] that the same integrals $I(z)$ and $\mathcal{T}(z)$ can be transformed into integrals $H(v)$ and $\mathcal{H}(v)$ to which the existence theorems in [10c] apply. Thus, I could obtain the expected existence theorems for the absolute minimum of $\mathcal{T}(z)$ for BV possibly discontinuous vector functions z, and of course $0 \le I(z) \le \mathcal{T}(z)$.

In [9b] I also studied a number of variants of the Serrin functional $\mathcal{T}(z)$ associated to the integral $I(z)$, namely, functionals $\mathcal{T}^*(z)$, $\mathcal{T}^{**}(z)$. I proved the needed properties of lower semicontinuity in the topology of L_1, and the basic relation $0 \le I(z) \le \mathcal{T}^{**}(z) \le \mathcal{T}^*(z) \le \mathcal{T}(z)$. It is clear that whenever we can prove that for the optimal solution z we have $I(z) = 0$, then z is a solution of the differential system $\sum_{j=1}^{\nu} [U_{ij}(t,z)]_{t_j} + V_i(t,z) = 0$, $i = 1, \ldots, m$, $t \in G$ (a.e).

By studying the Serrin integrals associated to a particular case of (1) with $m \ge 1$, $\nu = 1$, we prove in [11b] the existence of BV solutions $z(t,x), = (z_1, \ldots, z_m)$, $t \ge 0$, x scalar, of the Cauchy problem for systems of the form

$$z_{it} + (F_i(z))_x = 0,\ \ z_i(0,x) = w_i(x),\ \ i = 1, \ldots, m.$$

VI. RANKINE-HUGONIOT TYPE PROPERTIES IN TERMS OF THE CALCULUS OF VARIATIONS AND BV SOLUTIONS.

In [9c] I investigated in more details integrals $I(z)$ of the form

$$I(z) = \int_G |z_x + (F(z))_y| \, dxdy \tag{2}$$

and corresponding Serrin type functionals, say $0 \leq I(z) \leq \mathcal{T}^{**}(z) \leq \mathcal{T}^*(z) \leq \mathcal{T}(z)$.

For $m = 1$, $\nu = 1$, we are dealing with the original integral

$$I(z) = \int_G |z_x + (F(z))_y| \, dxdy, \; G \subset \mathbf{R}^2,$$
$$x, y, z, F \text{ scalars}. \tag{3}$$

If z has a line $\Gamma : y = \ell(x), a \leq x \leq b$, of class C^1 and of jump discontinuity, say

$$z_2(x) = z(x, \ell(x)+), \; z_1(x) = z(x, \ell(x)-), \; a \leq x \leq b,$$

then, under mild assumptions, the contribution of Γ on the value of the Serrin type functional $\mathcal{T}^*$ is ≥ 0, and such a contribution is zero if and only if

$$[z_2(x) - z_1(x)]\ell'(x) = F(z_2(x)) - F(z_1(x)), \; a \leq x \leq b,$$

along the line Γ (Cesari [9c]). This is the same relation which is well known for weak solutions of conservation laws (cf. Oleinik [18], Conway and Smoller [12]).

For $m = 1, \nu \geq 1$, we are dealing with the original integral

$$I(z) = \int_G |z_x + \sum_{j=1}^{\nu} (F_j(z))_{y_j}| \, dxdy, \; G \subset \mathbf{R}^{\nu+1}, \; dy = dy_1 \dots dy_\nu,$$
$$x, z \text{ scalar}, z(x,y) = z(x, y_1, \dots, y_\nu), \; F(z) = (F_1, \dots, F_\nu). \tag{4}$$

If $z(x,y)$ has a surface $\Gamma : x = L(y) = L(y_1, \dots, y_\nu)$, $y \in D$, of class C^1 and of jump discontinuity for z, say

$$z_2(y) = z(L(y)+, y), \; z_1(y) = z(L(y)-, y), \; y = (y_1, \dots, y_\nu) \in D,$$

then under mild assumptions the contribution of Γ on the value of the same Serrin type functional $\mathcal{T}^*$ is ≥ 0, and such a contribution is zero if and only if

$$z_2(y) - z_1(y) = \sum_{j=1}^{\nu} (L(y))_{y_j} [F_j(z_2(y)) - F_j(z_1(y))], \; y \in D,$$

on the surface Γ (Cesari [9c]).

For $m > 1$, $\nu = 1$ we are dealing with the original integral

$$I(u) = \int_G \sum_{i=1}^{m} |z_{ix} + (F_i(z))_y| \, dxdy, \; G \subset \mathbf{R}^2,$$

$$x, y, \text{ scalars}, \; z(x,y) = (z_1, \dots, z_m), \; (F(z) = (F_1, \dots, F_m), (5)$$

and in this situation we must use the Serrin type integral $\mathcal{T}^{**}$. Let us assume that for a given $i = 1, \dots, m$, the component $z_i(x,y)$ of z has a line $\Gamma : y = \ell(x)$, $a \le x \le b$, of class C^1 and of jump discontinuity for z_i, say

$$z_{i2}(x) = z_i(x, \ell(x)+), \; z_{i1}(x) = z_i(x, \ell(x)-), \; a \le x \le b,$$

while the remaining components $z_h(x,y)$, $h = 1, \dots, m$, $h \ne i$, are continuous in a neighborhood of Γ. In this situation, let us take

$$z^{i,2}(x) = (z_i(x, \ell(x)+); \; z_h(x, \ell(x)), \; h \ne i, \; h = 1, \dots, m),$$
$$z^{i,1}(x) = (z_i(x, \ell(x)-); \; z_h(x, \ell(x)), \; h \ne i, \; h = 1, \dots, m), \; a \le x \le b,$$

I proved in [9c], under mild assumptions, that the contribution of Γ on the value of the Serrin type integral $\mathcal{T}^{**}$ is ≥ 0, and such a contribution is zero if and only if

$$[z_{i2}(x) - z_{i1}(x)]\ell'(x) = F_i(z^{(i,2)}(x)) - F_i(z^{(i,1)}(x)), \; a \le x \le b,$$

along Γ (Cesari [9c]).

VII. THE WEIERSTRASS INTEGRAL W(z).

In [8ab] Cesari established a very general axiomatization concerning extensions of Burkill's integral on set functions. Namely, Cesari [8a] introduced a concept of quasi-additivity for set functions, guaranteeing the existence of a limit, now called the Burkill-Cesari integral. Namely, let A denote any topological space, let $\{I\}$ be a system of subsets I of A which we shall call intervals, and let $\phi(I)$, $I \in \{I\}$, be a given interval function $\phi(I) = (\phi_1, \dots, \phi_N)$. For any given net $(\mathcal{D}, \gg)$ of finite systems $D = (I_1, \dots, I_M)$ of nonoverlapping intervals $I_i \in \{I\}$, the limit

$$B(\phi) = \lim_{(\mathcal{D}, \gg)} \sum\nolimits_{i=1}^{N} \phi(I_i)$$

is called the Burkill-Cesari integral of the set function ϕ. Cesari [8a] proved that if ϕ is quasi additive, then $B(\phi)$ exists and is finite. About the non-linear integral $I = \int_T F(p,q)$ over a variety T, Cesari considered the set function $\Phi(I) = F(T(\omega(I)), \phi(I))$, where $\omega(I)$ is a choice function, i.e. $\omega(I) \in I$, and ϕ is a set function. Cesari proved [8a] that if T is any continuous parametric mapping and ϕ is quasi-additive and BV, then also Φ is quasi-additive and BV. In other words, the non-linear transformation F preserves quasi-additivity and bounded variation. Then the integral $W(z)$ is defined by the Burkill-Cesari process on the function Φ, and is thus defined as a Weierstrass-type integral

$$W(T,\phi) = \lim_{(\mathcal{D},\gg)} \sum_{i=1}^{N} \Phi(I_i).$$

Cesari further proved, under assumptions, that the Weierstrass integral $W(T,\phi)$ has a Lebesgue-Stieltjes integral representation

$$W(T,\phi) = \int_G F(T(s),\ (d\mu/d\|\mu\|)(s))d\mu$$

in terms of a vector measure μ related to ϕ, its total variation $\|\mu\|$, and the Radon-Nikodym derivative $d\mu/d\|\mu\|$ instead of Jacobians.

Later, many authors studied this integral, both in the parametric and in the non-parametric case, for curves and for varieties, and framed in this theory many of their properties (see [21] for a survey). Note that if F does not depend on the variety, i.e., it is of the type $F(q)$, then the sole concept of quasi-additivity permits the extension of $W(z)$ over BV curves and surfaces, not necessarily continuous nor Sobolev's.

In the last years Brandi and Salvadori [2defg] have extended the definition of $W(z)$ over BV curves or varieties, not necessarily continuous nor Sobolev's, for complete integrands $F(p,q)$.

First the term $T(\omega(I))$, in the definition of $\Phi(I)$ was replaced [2d] by a set function $P(I)$ whose values are in a metric space K, while $\phi(I)$ is a set function whose values are in a uniformly convex Banach space X and $F : K \times X \to E$, with E a real Banach space. In order to guarantee the existence of the integral $W(z)$ for BV transformations T, a condition on the pair of set functions (P,ϕ) was proposed in [2d], which is of the quasi-additivity-type, and was called Γ-quasi-additivity. This condition reduces to the quasi-additivity on ϕ when P is the usual set function $T(\omega(I))$ and T is continuous. In this new situation, Brandi and Salvadori proved that, if (P,ϕ) is Γ-quasi-additive and ϕ is BV, then still $\Phi(I) = F(P(I),\phi(I))$ is quasi-additive and BV. Thus the integral W is still defined by the Burkill-Cesari process on the set function Φ, and W is still a Weierstrass-type integral even for T only BV, possibly discontinuous.

Note that the new condition on (P,ϕ) is weaker than the couple of assumptions: continuity on T and quasi-additivity on ϕ. Moreover, it takes advantage of the power of the quasi-additivity-type properties to extend W over BV curves and varieties, for integrands of the type $F(p,q)$, both in the parametric and in the non-parametric case (see many applications in [2def]).

Even in this more general setting, the integral W admits of a Lebesgue-Stieltjes integral representation ([2d])

$$W(T,\phi) = \int_G F(T(s),(d\mu/d\|\mu\|)(s))d\|\mu\|,$$

in terms of a vectorial measure μ related to ϕ, its total variation $\|\mu\|$, and Radon-Nikodym derivative $d\mu/d\|\mu\|$, as in the previous work of Cesari [8b] in Euclidean spaces and in the successive extensions to abstract spaces, always for continuous varieties T (see [21] for a survey).

In the non-parametric case (see [2e]) Brandi and Salvadori transform the integral $I = \int_T f(t,p,q)$ into a suitable parametric integral in the manner of McShane, with integrand $F(t,p;\ell,q)$ defined by $F(t,p;\ell,q) = \ell f((t,p,q/\ell)$ for $\ell > 0$ and $F(t,p;0,q) = \lim_{\ell\to 0+} F(t,p;\ell,q)$. Then the set function Φ becomes $\Phi(I) = \lambda(I)f(p(I),\phi(I)/\lambda(I)) = F(p(I);\lambda(I),\phi(I))$, and the existence result is still given in terms of Γ-quasi-additivity. Now the representation of W in terms of Lebesgue-Stieltjes integral becomes

$$W(T,\phi) = \int_G f(\pi(s),(d(\nu,\mu)/d\|(\nu,\mu)\|)(s))\, d\|(\nu,\mu)\|,$$

where μ is the vectorial measure related to ϕ, ν is the real measure related to λ and $\|(\nu,\mu)\|$ is the total variation of the measure (ν,μ). Furthermore, in this non-parametric situation, Brandi and Salvadori proved a Tonelli-type inequality in [2e] relating $W(z)$ to a corresponding Lebesgue-Stieltjes integral, namely,

$$W(T,\phi) \geq \int_G f(T(s),(\delta\mu/\delta\nu)(s))\, d\nu,$$

where $\delta\mu/\delta\nu$ is a derivative of the Radon-Nikodym type, and the equality sign holds if and only if the set function ϕ is absolutely continuous with respect to the set function λ. If ϕ is absolutely continuous with respect to λ, then $\delta\mu/\delta\nu$ reduces to the usual Radon-Nikodym derivative $\partial\mu/\partial\nu$. In proving this last result, as in the proof of the representation theorem, Brandi and Salvadori used a connection between the Burkill-Cesari process and the convergence of martingales, a connection which was already made in previous papers (see [21] and also the quoted papers [2def]).

Finally, in [2f] Brandi and Salvadori dealt with the problem of the lower semicontinuity for the integral W, both in the parametric and in the non-parametric case. A first abstract lower semicontinuity theorem was proved in terms of a suitable global convergence on the sequence (P_n,ϕ_n), defined in the same spirit of the Γ-quasi-additivity and therefore again inspired to Cesari's concept of quasi-additivity. In a number of applications this convergence is implied by the L_1-convergence of equiBV varieties.

REFERENCES

1. G. Anzellotti and M. Giaquinta, "Funzioni BV e tracce", Rend. Sem. Mat. Padova 60 (1978), 1–21.
2. P. Brandi - A. Salvadori, (a) "Sull'integrale debole alla Burkill-Cesari", Atti Sem. Mat. Fis. Univ. Modena *27* (1978) 14–38; (b) "Existence, semi-continuity and representation for the integrals of the Calculus of Variations. The BV case", Atti Convegno celebrativo I centenario Circolo Matematico di Palermo (1984), 447–462; (c) "L'integrale del Calcolo delle Variazioni alla Weierstrass lungo curve BV e confronto con i funzionali integrali di Lebesgue e Serrin", Atti Sem. Mat. Fis. Univ. Modena *35* (1987); (d) "A quasi-additivity type condition and the integral over a BV variety", Pacific J. Math., to appear; (e) "On the non-parametric integral over a BV surface", Nonlinear Analysis, to appear; (f) "On the lower semicontinuity of certain integrals of the Calculus of Variations", J. Math. Anal. Appl., to appear;

(g) "On Weierstrass-type variational integrals over BV varieties", Rend. Accad. Naz. Lincei Roma, to appear.

3. J. C. Breckenridge, "Burkill-Cesari integrals of quasi additive interval functions", Pacific. J. Math. *37* (1971), 635–654.
4. F. Cafiero, "Criteri di compattezza per le successioni di funzioni generalmente a variazione limitata", Atti Accad. Naz. Lincei Roma *8* (1950), 305–310.
5. L. Cesari, "Sulle funzioni a variazione limitata", Annali Scuola Norm. Sup. Pisa (2) *5*, 1936, 299–313.
6. L. Cesari, (a) "Sulle funzioni di due variabili a variazione limitata e sulla convergenza delle serie doppie di Fourier", Rend. Sem. Mat. Univ. Roma *1*, 1937, 277–294; (b) "Sulle funzioni di più variabili a variazione limitata e sulla convergenza delle relative serie multiple di Fourier", Pontificia Accad. Scienze, Commentationes *3*, 1939, 171–197.
7. L. Cesari, (a) Surface Area, Princeton University Press, 1956; (b) Optimization-Theory and Applications, Springer Verlag, 1983.
8. L. Cesari, (a) "Quasi additive set functions and the concept of integral over a variety", Trans. Amer. Math. Soc. *102* (1962), 94–113; (b) "Extension problem for quasi additive set functions and Radon-Nikodym derivatives", Trans. Amer. Math. Soc. *102* (1962), 114–146.
9. L. Cesari, (a) "Existence of discontinuous absolute minima for certain multiple integrals without growth properties", Rend. Accad. Naz. Lincei Roma, to appear; (b) "Existence of discontinuous absolute minima for modified multiple integrals of the calculus of variations", to appear; (c) "Rankine-Hugoniot type properties in terms of the calculus of variations for BV solutions", to appear.
10. L. Cesari, P. Brandi, and A. Salvadori, (a) "Discontinuous solutions in problems of optimization", Annali Scuola Normale Sup. Pisa, to appear; (b) "Existence theorems concerning simple integrals of the calculus of variations for discontinuous solutions", Archive Rat. Mech. Anal. *98*, 1987, 307–328; (c) "Existence theorems for multiple integrals of the calculus of variations for discontinuous solutions", Annali Mat. Pura Appl., to appear. (d) "Seminormality conditions in the calculus of variations for BV solutions", to appear.
11. L. Cesari and P. Pucci, (a) "Remarks on discontinuous optimal solutions for simple integrals of the calculus of variations", Atti Sem. Mat.-Fis. Univ. Modena, to appear; (b) "Existence of BV discontinuous solutions of the Cauchy problem for conservation laws by variational argument", Atti Sem. Mat.-Fis. Univ. Modena, to appear.
12. E. Conway and J. Smoller, "Global solutions of the Cauchy problem for quasi linear first order equations in several space variables", Comm. Pure Appl. Math. *19*, 1966, 95–105.
13. C. M. Dafermos, "Generalized characteristics and the structure of solutions of hyperbolic conservation laws", Indiana Univ. Math. J. *26*, 1977, 1097–1119.
14. R. J. Di Perna, "Singularities of solutions of nonlinear hyperbolic sytems of conservation laws", Archive Rat. Mech. Anal. *60*, 1974, 75–100.
15. W. H. Fleming, "Functions with generalized gradient and generalized surfaces", Annali Mat. Pura Appl. *44*, 1957, 93–103.
16. E. Gagliardo, "Proprietà di alcune classi di funzioni di più variabili", Ricerche Mat. *7*, 1959, 24–51.
17. K. Krickeberg, "Distributionen, Funktionen beschränkter Variation und Lebesguescher Inhalt nichtparametrischer Flächen", Annali Mat. Pura Appl. *44*, 1957, 105–133.
18. O. A. Oleinik, "On discontinuous solutions of nonlinear differential equations", Uspekhi Mat. Nauk *12*, 1957, 3–73. English translation, Amer. Math. Soc. Transl. (2) *26*, 95–172.

19. A. Salvadori, "Theorems of lower semicontinuity", to appear.
20. J. Serrin, (a) "On the definition and properties of certain variational integrals", Trans. Amer. Math. Soc. *101*, 1961, 139–167; (b) "On the differentiability of functions of several variables", Archive Rat. Mech. Anal. *7*, 1961, 359–372.
21. C. Vinti, "Non linear integration and Weierstrass integral over a manifold. Connection with theorems on martingales", Journ. of Optimization Theory Applications *41*, 1983, 213–237.
22. A. L. Volpert, "The space BV and quasi linear equations", Mat. Sb. *73*, 1967, 225–267.
23. G. Warner, (a) "The Burkill-Cesari integral", Duke Math. J. *35*, 1968, 61–78; (b) "The generalized Weierstrass-type integral $\int f(\xi, \phi)$." Annali Scuola Norm. Sup. Pisa *22*, 1968, 163–192.

Minimizing sequences for nonconvex functionals, phase transitions and singular perturbations

Stefan Müller
Department of Mathematics
Heriot-Watt University
Edinburgh EH14 4AS
Scotland, UK

Abstract

We study minimizers of the singularly perturbed functional $I^{\epsilon}(u) = \int_0^1 \{\epsilon^6 u_{xx}^2 + (u_x^2 - 1)^2 + u^2\}\, dx$, subject to $u(0) = u(1) = 0$. For $\epsilon = 0$ no minimizers exist and we show that for $\epsilon > 0$, small, the minimizer is nearly periodic with period proportional to ϵ. Connections with solid-solid phase transitions in crystals are indicated.

1 Introduction

Continuum models based on the minimization of nonconvex functionals have been used to model a variety of phase transitions. A problem often encountered is that the functional has many minimizers. A *selection principle* is needed to choose the physically relevant ones. The present work, motivated by a model for solid-solid phase transitions in crystals, addresses the situation where the underlying functional has no minimizers and analyses a selection criterion for *minimizing sequences.*

For illustration consider first an example leading to many minimizers. The free energy of a van der Waals gas (at constant temperature) confined to a container $\Omega \subset R^n$ is given by

$$F(v) = \int_{\Omega} f(v(x))dx, \tag{1}$$

where v is the density of the gas. We seek to minimize F subject to the constraint that the total mass is given, *i.e.*

$$\int_{\Omega} v\, dx = m = \lambda \operatorname{meas} \Omega. \tag{2}$$

Coexistence of different phases can occur if f is a nonconvex function. Replacing f by $f(v) - \mu_1 v - \mu_2$ (which in view of (2) does not change the minimizers of F) we may assume that f has exactly two minima, at a and b, with minimum value 0. For $\lambda \in (a, b)$, v is a minimizer of F if and only if, for some $A \subset \Omega$ with $\operatorname{meas} A = \frac{b-\lambda}{b-a} \operatorname{meas} \Omega$,

$$v(x) = \begin{cases} a & \text{for} x \in A \\ b & \text{for} x \in \Omega \setminus A \end{cases} \tag{3}$$

If we regard $v = a$ and $v = b$ as two phases then only the *proportion* of each phase is determined by minimizing F. No information is obtained about the geometrical arrangement of the phases. This leads to the question whether some of the minimizers are — in some physical sense — preferred.

The approach taken by van der Waals [vdW 93] and rediscovered by Cahn and Hilliard [CH 58] is to consider a modified energy functional which also depends on the density gradient and thus penalizes sharp transitions in density. The problem becomes then to minimize

$$F^\epsilon(v) = \int_\Omega \{\epsilon^2 \mid \nabla v \mid^2 + f(v)\} dx ,$$

subject to (2).

If Ω is an interval, Carr, Gurtin and Slemrod [CGS 84] showed that for sufficiently small ϵ the minimizer v^ϵ has to be a monotone function. Letting $\epsilon \to 0$ one obtains a minimizer $\overline{v}$ of F such that the sets A and $\Omega \setminus A$ in (3) are intervals which meet in exactly one point. This result was generalized by Modica [Mo 87] (see also his article in the present volume) and Kohn and Sternberg [KS 88] to $\Omega \subset R^n$. These authors show that for the limiting minimizer $\overline{v}$ the interface between A and $\Omega \setminus A$ is a minimal surface (subject to the volume constraint). Thus the study of the singularly perturbed functional F^ϵ leads, in this case, to a sensible selection of 'preferred' minimizers of F.

The present work was motivated by an attempt to use the singular perturbation approach in the context of solid-solid phase transitions in crystals. Ball and James [BJ 87], [BJ 89] used a continuum model to study these phase transitions which is based on the minimization of the elastic free energy

$$E(u) = \int_\Omega W(\nabla u)\, dx ,$$

where $\Omega \subset R^3$ denotes the reference configuration of the crystal, $u : \Omega \to R^3$ its deformation and W the stored-energy density. It turns out that in general the infimum of E is *not* attained. In a particular example one obtains minimizing sequences $u^{(n)}$ such that $\nabla u^{(n)}$ essentially only takes two values F_1, F_2 on sets A_1, A_2. As $E(u^{(n)})$ approaches the infimum, the layering of A_1 and A_2 becomes increasingly finer while the relative *proportion* of the two 'phases', *i.e.* the ratio $\text{meas}A_1/\text{meas}A_2$ approaches a limit. This, however, still leaves a great degree of freedom as regards the detailed arrangement of A_1 and A_2. There exist, *e.g.* both periodic and nonperiodic minimizing sequences. In order to select 'preferred' minimizing sequences one would like to study the minimizers of

$$E^\epsilon(u) = \int_\Omega \{\epsilon^2 \mid \nabla^2 u \mid^2 + W(\nabla u)\}\, dx ,$$

and pass to the limit $\epsilon \to 0$. This, however, appears to be a rather difficult problem and in the present work we confine ourselves to a one-dimensional problem instead.

For $u : [0,1] \to R$ consider

$$I(u) = \int_0^1 \{(u_x^2 - 1)^2 + u^2\}\, dx .$$

Under the boundary conditions $u(0) = u(1) = 0$ the infimum of I is not attained. If $u^{(n)}$ is a minimizing sequence then $u_x^{(n)}$ has to oscillate increasingly faster between -1

and 1. As in the previous example there is a variety of such sequences. It turns out, however, that the minimizing sequences obtained via singular perturbations exhibit an (approximately) periodic pattern. More precisely we have

Theorem 1.1 *Let*

$$I^\epsilon(u) = \int_0^1 \{\epsilon^6 u_{xx}^2 + (u_x^2 - 1)^2 + u^2\}\, dx\,, \tag{4}$$

let ϵ be sufficiently small and let u be a minimizer of I^ϵ, subject to

$$u(0) = u(1) = 0\,. \tag{5}$$

Then u_x has a finite number of zeros $0 < x_1 < \ldots < x_{N^\epsilon} < 1$, $N^\epsilon = \epsilon^{-1} L_0^{-1} + \mathcal{O}(1)$, and

$$x_{i+1} - x_i = \epsilon L_0 + \mathcal{O}(\epsilon^2). \tag{6}$$

Moreover

$$u_x = \pm 1 + \mathcal{O}(\epsilon^2), \tag{7}$$

provided that $| x - x_i | \geq C\epsilon^3 \ln \epsilon^{-1}$, for all $i = 1, \ldots, N_\epsilon$. Finally

$$\min I^\epsilon = \epsilon^2 E_0 + \mathcal{O}(\epsilon^4). \tag{8}$$

Thus for the minimizers u^ϵ of I^ϵ, u_x^ϵ approaches a step function with approximately equal steps as $\epsilon \to 0$. We have used the notation

$$A_0 = 2\int_{-1}^{1} (1 - z^2)\, dz = \frac{8}{3}, \tag{9}$$

$$L_0 = (6A_0)^{1/3} = 2\sqrt[3]{2}, \tag{10}$$

$$\begin{aligned} E_0 &= \min_d (A_0 d^{-1} + \frac{1}{12} d^2) \\ &= A_0 L_0^{-1} + \frac{1}{12} L_0^2 = \sqrt[3]{4}. \end{aligned} \tag{11}$$

Remarks. 1. Note the unusual scaling ϵ^6 of the singular perturbation. This was merely incorporated to avoid writing fractional powers in the sequel.

2. The scaling law for N^ϵ and I^ϵ have been predicted by Tartar [Ta 87] on the base of formal asymptotic expansions.

3. Similar results can be proved if $(u_x^2 - 1)^2$ is replaced by a more general double well potential.

The proof of the theorem comprises two steps. First we derive very precise upper and lower bounds for $\min I^\epsilon$. These imply pointwise bounds on u^ϵ, uniformly in ϵ. Equipped with this additional information we employ in a second step the Euler-Lagrange equations to estimate the location of the zeros of u_x^ϵ. The first step is carried out in section 2, the second in section 3.

2 Variational estimates

In view of (6) it seems reasonable to rescale x and we define

$$v(x) = \epsilon^{-1}u(\epsilon x). \tag{12}$$

This gives

$$v_x(x) = u_x(\epsilon x), \quad v_{xx}(x) = \epsilon u_{xx}(\epsilon x),$$

and

$$\begin{aligned} I^\epsilon(u) &= \int_0^1 \{\epsilon^6 u_{xx}^2 + (u_x^2 - 1)^2 + u^2\} dx \\ &= \int_0^{1/\epsilon} \{\epsilon^4 v_{xx}^2 + (v_x^2 - 1)^2 + \epsilon^2 v^2\} \epsilon \, dx \\ &= \epsilon^3 \int_0^{1/\epsilon} \{\epsilon^2 v_{xx}^2 + \epsilon^{-2}(v_x^2 - 1)^2 + v^2\} dx. \end{aligned} \tag{13}$$

Let

$$J^\epsilon(v; a, b) = \int_a^b \{\epsilon^2 v_{xx}^2 + \epsilon^{-2}(v_x^2 - 1)^2 + v^2\} dx. \tag{14}$$

Then

$$J^\epsilon(v) = J^\epsilon(v; 0, \epsilon^{-1}) = \epsilon^3 I^\epsilon(u). \tag{15}$$

The main result of this section is

Theorem 2.1 *Assume that ϵ is sufficiently small and that $b - a \leq \epsilon^{-1}$. Then there exists a constant C such that*

$$\min\{J^\epsilon(v; a, b) \mid v \in H^2(a, b), v(a) = v(b) = 0\} \leq E_0(b - a) + C, \tag{16}$$

$$\min\{J^\epsilon(v; a, b) \mid v \in H^2(a, b)\} \geq E_0(b - a) - C, \tag{17}$$

where E_0 is given by (11).

Remark. The proof that the minimum of J^ϵ is attained is standard since the integrand is convex and coercive in the highest derivatives.

Theorem 2.1 immediately leads to pointwise estimates for v.

Corollary 2.2 *Let v be a minimizer of J^ϵ subject to $v(0) = v(\epsilon) = 0$. Then we have, for every interval $(a, b) \subset (0, 1/\epsilon)$,*

$$J^\epsilon(v; a, b) \leq E_0(b - a) + C. \tag{18}$$

Moreover

$$\sup_{(0,1/\epsilon)} \mid v \mid + \mid v_x \mid \leq C. \tag{19}$$

Remark. Note that (12), (19) imply that $| u | \leq C\epsilon$, if u is a minimizer of the original functional I^ϵ.

Proof of Corollary 2.2 . The first assertion is immediate. Observe that $J^\epsilon(v;a,b) = J^\epsilon(v;0,1/\epsilon) - J^\epsilon(v;0,a) - J^\epsilon(v;b,1/\epsilon)$ and apply Theorem 2.1. As for the second assertion we first show $\sup | v_x | \leq C$. Let $M = \sup |v_x| - 1$, assume without loss of generality $M \geq 4$ and choose x_0 such that $v_x(x_0) = M \geq 4$. Let $(a,b) \subset (0,1/\epsilon)$ be the maximal interval containing x_0 on which $v_x > 2$ (note that v_x is continuous since $v \in H^2$). Clearly $(a,b) \neq (0,1/\epsilon)$ since v is minimizing. Therefore $v_x = 2$ at $x = a$ or $x = b$. Assume the former, then, by Lemma 2.3 below,

$$\int_a^{x_0} \{\epsilon^2 v_{xx}^2 + \epsilon^{-2}(v_x^2 - 1)^2\} dx \geq 2 \int_2^M (z^2 - 1) dz \geq \frac{2}{3} M^3 - C.$$

Moreover, by splitting v in its positive and its negative part, we see that $v_x \geq 2$ implies

$$\int_a^b v^2 dx \geq \int_0^{(b-a)/2} (2z)^2 dz = \frac{1}{6}(b-a)^3,$$

and hence

$$E_0(b-a) + C \geq J^\epsilon(v;a,b) \geq \frac{2}{3} M^3 + \frac{1}{6}(b-a)^3 - C,$$

so that $M \leq C$, uniformly in ϵ. Finally, for $a \in (0,1/\epsilon)$, observe that

$$| v(a) - \int_a^{a+1} v\, dx | \leq \sup | v_x | \leq C,$$

and

$$| \int_a^{a+1} v\, dx | \leq \left\{ \int_a^{a+1} v^2\, dx \right\}^{1/2} \leq (E_0 + C)^{1/2},$$

which imply that $\sup | v | \leq C$.

The proof of Theorem 2.1 is based on the following observations. The minimizer of $J^\epsilon(.;a,b)$ will satisfy $v_x \sim \pm 1$ except on small transition layers where v_x changes sign. The term $\epsilon^2 v_{xx}^2 + \epsilon^{-2}(v_x^2 - 1)^2$ will only be large in these transition layers and the integral from a to b of this term will increase with the number of transition layers but will be more or less independent of their mutual distance. On the other hand, v^2 will become large if v_x is constant over a long interval, *i.e.* if the distance between two transition layers becomes large. The combination of these two effects will force the transition layers to be more or less equidistant, and there is some optimal value L_0 for their mutual distance. We begin with a lemma due to Modica [Mo 87] which gives a lower bound for the energy associated with one transition layer.

Lemma 2.3 *Let*

$$H(z) = \int_0^z | 1 - p^2 | \, dp \tag{20}$$

and let $v \in H^2(a,b)$. *Then*

$$\int_a^b \{\epsilon^2 v_{xx}^2 + \epsilon^{-2}(v_x^2 - 1)^2\}\, dx \geq 2 \, | H(v_x)(b) - H(v_x)(a) | \tag{21}$$

Proof. We have

$$\begin{aligned}
&\int_a^b \epsilon^2 v_{xx}^2 + \epsilon^{-2}(v_x^2-1)^2\,dx \\
\geq\ &\int_a^b 2\mid v_x^2-1\mid\mid v_{xx}\mid dx \\
\geq\ & 2\int_a^b \mid H(v_x)_x\mid dx \geq 2\mid H(v_x)(b)-H(v_x)(a)\mid
\end{aligned}$$

Proof of Theorem 2.1 (i) (upper bound). We chose a testfunction v with N equidistant transition layers. Specifically let $x_i = a + \frac{2i-1}{2N}(b-a)$, $i=1,\ldots,N$, let

$$w(x) = \begin{cases} -\tanh \epsilon^{-2}(x-x_i), & \mid x - x_i\mid \leq \frac{1}{2N},\ i \text{ odd}, \\ \tanh \epsilon^{-2}(x-x_i), & \mid x - x_i\mid \leq \frac{1}{2N},\ i \text{ even}, \end{cases}$$

and let

$$v(x) = \int_a^x w(\xi)\,d\xi,$$

(*cf.* [Ta 87]). We clearly have $v(a) = v(b) = 0$. Moreover on $(x_i - \frac{b-a}{2N}; x_i + \frac{b-a}{2N}) = (x^-; x^+)$ we have

$$\epsilon^2 v_{xx}^2 = \epsilon^{-2}(1-v_x^2)^2$$

so that the estimate in Lemma 2.3 becomes sharp (this was in fact the rationale for the choice of w) and hence

$$\begin{aligned}
&\int_{x^-}^{x^+} \{\epsilon^2 v_{xx}^2 + \epsilon^{-2}(1-v_x^2)^2\}\,dx \\
&= 2\mid H(v_x)(x^+) - H(v_x)(x^-)\mid \\
&\leq 2(H(1)-H(-1)) = A_0.
\end{aligned} \tag{22}$$

(*cf.* (9)). Moreover $v(x_i + \frac{b-a}{2N}) = 0$, and hence

$$\begin{aligned}
\int_{x^-}^{x^+} v^2\,dx &= 2\int_{x_i}^{x^+} v^2\,dx \leq 2\int_{x_i}^{x^+}(x^+ - x_i)^2\,dx \\
&= \frac{1}{12}\left(\frac{b-a}{N}\right)^3
\end{aligned} \tag{23}$$

It follows that

$$\begin{aligned}
J^\epsilon(v;a,b) &\leq N(A_0 + \frac{1}{12}\frac{(b-a)^3}{N^3}) \\
&= (b-a)(A_0 d^{-1} + \frac{1}{12}d^2),
\end{aligned} \tag{24}$$

where $d = (b-a)/N$. If we could chose d arbitrarily, the left hand side of (24) would be minimized by $d = L_0$, with minimum value E_0 (see (11)). Choosing N as the smallest integer larger than $(b-a)/d$, we find that

$$J^\epsilon(v;a,b) \leq E_0(b-a) + A_0,$$

which proves (16). Expanding the function $d \mapsto A_0 d^{-1} + \frac{1}{12}d^2$ around L_0, we see that in fact

$$J^{\epsilon}(v;a,b) \leq (b-a)(E_0 + C(b-a)^{-2}). \tag{25}$$

Proof of Theorem 2.1 (ii) (lower bound). For the function v used to establish the upper bound we found that every transition layer (*i.e.* every interval where v_x changes from $\sim +1$ to ~ -1 or vice versa) contributes an amount A_0 to the energy, while an interval of length d between two consecutive transition layers contributes $\frac{1}{12}d^3$. Now we want to show that (up to order ϵ) we cannot do better. The following definition will be useful.

Definition 2.4 *Let $v \in H^2(a,b), \delta \in (0,1)$. An interval (x^-, x^+) is called a* **δ-transition layer** *for v if*

$$\begin{array}{lcll} v_x(x^-) & = & -1+\delta & (1-\delta, \textit{resp.}) \\ v_x(x^+) & = & 1-\delta & (-1+\delta, \textit{resp.}) \\ v_x(x) & \in & (-1+\delta, 1-\delta) & \textit{for } x \in (x^-, x^+) \end{array}$$

Lemma 2.5 *Let ϵ be sufficiently small, let $\delta = \epsilon^{1/2}$ and let $v \in H^2(a,b)$. If (x^-, x^+), (y^-, y^+) are two consecutive δ-transition layers for v then, for all $y \in (x^+, y^-)$,*

$$\int_{x^-}^{y} \{\epsilon^2 v_{xx}^2 + \epsilon^{-2}(v_x^2 - 1)^2 + v^2\}\, dx \geq (1 - C\epsilon)E_0(y - x^-), \tag{26}$$

where E_0 is given by (11). If (x,y) is an interval which does not intersect any δ-transition layer, then

$$\int_{x}^{y} \{\epsilon^2 v_{xx}^2 + \epsilon^{-2}(v_x^2 - 1)^2 + v^2\}\, dx \geq (1 - C\epsilon)E_0(y - x) - A_0, \tag{27}$$

where A_0 is given by (9).

Proof. We only show the first assertion as the proof of the second is similar. Fix $y \in (x^+, y^-)$ and let $d = y - x^-$. By Lemma 2.3 and Definition 2.4 we immediately obtain

$$\begin{aligned} \int_{x^-}^{y} \{\epsilon^2 v_{xx}^2 + \epsilon^{-2}(v_x^2 - 1)^2\}\, dx &\geq \int_{x^-}^{x^+} \{\epsilon^2 v_{xx}^2 + \epsilon^{-2}(v_x^2 - 1)^2\}\, dx \\ &\geq 2(H(1-\delta) - H(-1+\delta)) \\ &\geq A_0 - c\delta^2 \geq (1 - C\epsilon)A_0 \end{aligned} \tag{28}$$

Moreover we may assume

$$\int_{x^-}^{y} (v_x^2 - 1)^2\, dx \leq E_0 d\epsilon^2, \tag{29}$$

as otherwise there is nothing to show. Letting $v(x) = v(x^-) + x - x^- + w(x) = v_1(x) + w(x)$ we have

$$\int_{x^-}^{y} v_1^2\, dx \geq \frac{d^3}{12}. \tag{30}$$

Now $w_x = v_x - 1$. Since (x^-, y) contains only one transition layer, either $v_x \leq 1 - \delta$ or $v_x \geq -1 + \delta$ on the whole interval. Assume that the later holds. We find $\mid w_x \mid\leq$

$v_x^2 - 1$ if $v_x \geq 0$, $| w_x | \leq \delta^{-2}(v_x^2 - 1)^2$ if $-1 + \delta \leq v_x \leq 0$. Therefore, using (29), it follows that

$$\begin{aligned} | w(x) | &\leq \int_{x^-}^{y} | w_x | \, dx \leq \int_{x^-}^{y} \{(v_x^2 - 1) + \delta^{-2}(v_x^2 - 1)^2\} \, dx \\ &\leq d^{1/2}(E_0 d\epsilon^2)^{1/2} + \delta^{-2} E_0 d\epsilon^2 \\ &= (E_0^{1/2} + E_0) d\epsilon, \end{aligned}$$

and

$$\int_{x^-}^{y} | w |^2 \, dx \leq C\epsilon^2 d^3,$$

which, together with (30), implies

$$\int_{x^-}^{y} | v |^2 \, dx \geq (1 - C\epsilon)\frac{d^3}{12}. \tag{31}$$

Now (26) follows from (28) and (31) since $A_0 + d^3/12 \geq E_0 d$ by definition of E_0.

Proof of Theorem 2.1 (ii) (continued). We only need to show that $J^\epsilon(v; a, b) \geq E_0(b-a) - C$, whenever v satisfies $J^\epsilon(v; a, b) \leq E_0(b-a)$. Fix one such v. By Lemma 2.3, v can only contain a finite number of δ- transition layers (we chose $\delta = \epsilon^{1/2}$) (x_i^-, x_i^+), $i = 1, \ldots, N$. Applying (26) to (x_i^-, x_{i+1}^-) (where $x_{N+1}^- = b$) and (27) to (a, x_1^-) it follows that

$$\begin{aligned} J^\epsilon(v; a, b) &\geq (1 - C\epsilon)E_0(b - a) - A_0 \\ &\geq E_0(b - a) - C, \end{aligned}$$

since by assumption $b - a \leq \epsilon^{-1}$. This finishes the proof of Theorem 2.1.

3 Analysis of the Euler-Lagrange equation

Let $v \in H^2(0, \epsilon^{-1})$ be a minimizer of

$$J^\epsilon(v) = \int_0^{1/\epsilon} \{\epsilon^2 v_{xx}^2 + \epsilon^{-2}(v_x^2 - 1)^2 + v^2\} \, dx, \tag{32}$$

subject to

$$v(0) = v(\epsilon^{-1}) = 0. \tag{33}$$

Then v is in fact smooth and satisfies the Euler-Lagrange equation

$$\epsilon^4 v_{xxxx} - 2(v_x^3 - v_x)_x + \epsilon^2 v = 0, \tag{34}$$

together with the boundary conditions

$$v = v_{xx} = 0 \text{ at } x = 0, \epsilon^{-1}. \tag{35}$$

The proof of Theorem 1.1 will be based on a careful analysis of (34), using in particular the a priori estimates of Corollary 2.2. The main idea is to consider the term $\epsilon^2 v$ as a slowly varying perturbation. We will show that transitions layers, *i.e.* intervals where v_x

changes sign, can only occur near a point x_0 if $\pm v(x_0)$ is close to a certain constant (see Lemma 3.5). This in turn will imply that the transition layers have to be (approximately) equidistant. Two important steps towards these results are contained in Lemmas 3.3 and 3.4, respectively. The former states that if $v_x(x_0)$ is not close to ± 1 a transition layer must occur near x_0, while the latter, based on linearization, ensures that if v_x is close to ± 1 on a sufficiently long interval one can obtain very sharp estimates for v_x on a slightly smaller interval. In the following v *will always denote a (global) minimizer of* J^ϵ (subject to (33)), unless otherwise stated.

We begin with a very modest result ensuring that every sufficiently long interval contains a subinterval of fixed length on which v_x is close to 1.

Lemma 3.1 *There exist constants L and B such that for every $\delta \in (0, 1/2)$, every $\epsilon \leq B\delta$, every $\Delta \leq B\delta^2$ and every interval $(a, a+L) \subset (0, \epsilon^{-1})$ there exits a subinterval $(a', a'+\Delta) \subset (a, a+L)$ on which $1-\delta < v_x < 1+\delta$. The same result holds with v_x replaced by $-v_x$.*

Proof. We give a sketch of the argument which is based on (18). Let $M = \{x \in (a, a+L) \mid v_x \in (1-\delta, 1+\delta)\}$. M is a union of open intervals which we divide into two groups. The first group contains those intervals for which $\mid v_x - 1 \mid \geq \delta/2$ on the whole interval; denote their union by M_1. The second group contains the remaining intervals, their union is denoted by M_2. If J is an interval from the second group, v_x must take the values $1 \pm \delta$ and $1 \pm \delta/2$ on $\bar{J}$. Thus, by Lemma 2.3 and (18) there can be at most $C\delta^{-2}(L+1)$ such intervals. If the lemma was false we therefore had $\text{meas} M_2 \leq C\delta^{-2}(L+1)\Delta$. On M_1 we have $W(x) \geq c\delta^2$, and hence, again by (18), $\text{meas} M_1 \leq C\delta^{-2}\epsilon^2(L+1)$. Similarly $\text{meas} M_3 \leq C\delta^{-2}\epsilon^2(L+1)$, where $M_3 = \{x \in (a, a+L) \mid v_x \notin (-1-\delta, -1+\delta) \cup (1-\delta, 1+\delta)\}$. Now $v_x \in (-1-\delta, -1+\delta)$ on $(a, a+L) \setminus (M_1 \cup M_2 \cup M_3)$, and hence, by the bound on v_x in (19), $\int_a^{a+L} v_x \, dx \leq -L/4$ for a suitable choice of the constant B. For large enough L this contradicts the uniform bound (19) on v and thus the lemma is proved.

It will be useful to consider the integrated form of (34). Letting

$$V(x) = \int_0^x v(\tau) \, d\tau, \tag{36}$$

we obtain

$$\epsilon^2 v_{xxx} - 2\epsilon^{-2}(v_x^3 - v_x) + V = C_1^\epsilon. \tag{37}$$

Multiplying by v_{xx} and integrating we deduce after various integrations by parts

$$\epsilon^2 v_{xx}^2 - \epsilon^{-2}(v_x^2 - 1)^2 = -2(V - C_1^\epsilon)v_x + v^2 + C_2^\epsilon \tag{38}$$

Lemma 3.2 *There exists a constant C (independent of ϵ) such that for all $x \in (0, \epsilon^{-1})$*

$$\mid C_1^\epsilon \mid + \mid C_2^\epsilon \mid + \mid V(x) \mid \leq C. \tag{39}$$

Proof. We first show $\mid C_1^\epsilon \mid \leq C$. By Lemma 3.1 there exists a constant C and intervals $I_1 = (a', a'+C^{-1}\epsilon^2)$, $I_2 = (b', b'+C^{-1}\epsilon^2)$ such that $I_i \subset (0, L)$ and $\mid v_x - 1 \mid \leq C\epsilon$ on I_1, $\mid v_x + 1 \mid \leq C\epsilon$ on I_2. By the Mean Value Theorem there exist $a \in I_1$, $b \in I_2$ with

$| v_{xx} | \le C^{-1}\epsilon^{-1}$ at a, b. Applying (38) at a, b, taking the difference and using (19) we see that $| C_1^\epsilon | \le C$, since $| V(b) - V(a) | \le L \sup | v | \le C$. Taking sums leads to $| C_2^\epsilon | \le C$.

To show that $| V(x) | \le C$ we choose $b \in (x, x+L)$ such that $| v_{xx}(b) | \le C\epsilon^{-1}$, $| v_x(b) - 1 | \le C\epsilon$, use (38) for a, b and take the difference.

Next we show that the transition regions where v_x changes sign have to be rather thin.

Lemma 3.3 *Let $\delta \ge \overline{C}\epsilon$, where $\overline{C}$ is a sufficiently large constant.*

1. *If x_0 is a zero of v_x, then there exist x_1, x_2 such that*

$$-v_x(x_1) = v_x(x_2) = 1 - \delta,$$

$$\frac{1}{4}\epsilon^2 \ln 2\delta^{-1} \le | x_i - x_0 | \le \epsilon^2 \ln 2\delta^{-1}.$$

2. *Conversely, if $\pm v_x(x_1) = 1 - \delta$, there exits a zero x_0 of v_x such that $| x_1 - x_0 | \le \epsilon^2 \ln 2\delta^{-1}$.*

3. *If $v_x(x_0) \ge 1 + \delta$ (resp. $v_x(x_0) \le -1 - \delta$), then $v_x \ge 1 + \delta$ (resp. $v_x \le -1 - \delta$), either on $(0, x_0)$ or on (x_0, ϵ^{-1}).*

Proof. We only show 1., the other proofs being similar. Let $w = v_x$. From (38), (19) and Lemma 3.2 we obtain

$$\epsilon^4 w_x^2 \ge (w^2 - 1)^2 - C\epsilon^2.$$

In particular w_x cannot change sign unless w is close to ± 1. By choosing $\overline{C}$ large enough we may achieve that $(w^2 - 1)^2 - C\epsilon^2 \ge \frac{1}{4}(w^2 - 1)^2 \ge C\delta^2$ for $w \in [-1 + \delta, 1 - \delta]$. In particular we can invert $x \mapsto w(x)$ until $\pm w$ reaches $1 - \delta$ and, assuming for convenience $w_x > 0$, we find

$$\begin{aligned} x_2 - x_0 &= \int_0^{1-\delta} \frac{dw}{w_x(w)} \le 2\epsilon^2 \int_0^{1-\delta} \frac{dw}{1 - w^2} \\ &= 2\epsilon^2 \operatorname{arctanh}(1 - \delta) \le \epsilon^2 \ln 2\delta^{-1}. \end{aligned} \tag{40}$$

The lower bound for $x_2 - x_0$ and the bounds for $x_1 - x_0$ are obtained similarly.

As a consequence of Lemma 3.3 we establish that

$$| v_x | \le 1 + C\epsilon,$$

for a suitable constant C. From (38), (35), (39) and (19) we obtain that, for $x = 0$ and $x = \epsilon^{-1}$, $(v_x^2 - 1)^2 \le C\epsilon^2$, and hence $| v_x | \le 1 + C\epsilon$. By Lemma 3.3 3. the same holds for all $x \in (0, \epsilon^{-1})$.

The interval $(0, \epsilon^{-1})$ may thus be divided into regions where $\pm v_x \in (1 - \delta, 1 + \delta)$, $\delta = C\epsilon$, and transition layers where $v_x \in (-1 + \delta, 1 - \delta)$. The latter are very thin by Lemma 3.3. If the former are sufficiently long sharper estimates on v_x can be obtained. We have

Lemma 3.4 *If* $| v_x - 1 | \leq C\epsilon$ *on* (a, b) *then*

$$| v_x - 1 | + \epsilon^2 | v_{xx} | \leq C\epsilon^2 \tag{41}$$

on $(a + \tau, b - \tau)$, *where* $\tau = 2\epsilon^2 \ln 2\epsilon^{-1}$.

If $v_x > 0$ *on* (a, b) *then*

$$| v_x - 1 | + \epsilon^2 | v_{xx} | \leq C\epsilon^2 \tag{42}$$

on $(a + 2\tau, b - 2\tau)$. *The same estimates hold if* v_x *is replaced by* $-v_x$.

Proof. Note that the second assertion is an immediate consequence of the first and Lemma 3.3. As for the first assertion we show first $| v_x - 1 | \leq C\epsilon^2$ on $(a + \tau, b - \tau)$. Expanding $2(z^3 - z)$ around $z = 1$ we obtain $2(z^3 - z) = 4(z - 1) + \mathcal{O}(| z - 1 |^2)$ so that (37) becomes

$$\epsilon^4 w_{xx} - 4(w - 1) = \epsilon^2 (C_1^\epsilon - V) + \mathcal{O}(| w - 1 |^2) = \mathcal{O}(\epsilon^2), \tag{43}$$

where we wrote $w = v_x$. Let w^0 be the solution of $\epsilon^4 w^0_{xx} - 4(w^0 - 1) = 0$, subject to $w^0(a) = w(a), w^0(b) = w(b)$. Then $w - w^0$ has zero boundary values on (a, b) and satisfies

$$\epsilon^4 (w - w^0)_{xx} - 4(w - w^0) = \mathcal{O}(\epsilon^2).$$

By the maximum priciple, $| w - w^0 | \leq C\epsilon^2$ on (a, b). It is easily checked that

$$\begin{aligned} w^0(x) \quad = \quad & 1 + \frac{\Delta(a) + \Delta(b)}{2} \frac{\cosh\{(x - \bar{x})/2\epsilon^2\}}{\cosh(d/2\epsilon^2)} \\ & + \frac{\Delta(a) - \Delta(b)}{2} \frac{\sinh\{(x - \bar{x})/2\epsilon^2\}}{\sinh(d/2\epsilon^2)}, \end{aligned}$$

where $\Delta = v_x - 1, \bar{x} = (a+b)/2, d = (b-a)/2$. The desired estimate for $| v_x - 1 |$ follows. Moreover (43) implies that $| v_{xxx} | = | w_{xx} | \leq C\epsilon^{-2}$ on $(a + \tau, b - \tau)$. Using the Mean Value Theorem on an interval of length ϵ^2 together with the estimate for v_x we obtain the estimate for v_{xx}.

The next lemma provides the key estimate for the proof of Theorem 1.1.

Lemma 3.5 *Assume that any two zeros of* v_x *are at least a distance* $4\tau = 8\epsilon^2 \ln 2\epsilon^{-1}$ *apart. Then at every zero* x *of* v_x *we have*

$$| v^2(x) + C_2^\epsilon | + | V(x) - C_1^\epsilon | \leq C\epsilon^2. \tag{44}$$

Moreover $C_2^\epsilon \geq c > 0$ *(uniformly in* ϵ*), and any two consecutive zeros* x, y *of* v_x *satisfy*

$$| y - x - 2(-C_2^\epsilon)^{1/2} | \leq C\epsilon^2, \tag{45}$$

while the first and the last zero of v_x *satisfy*

$$| x_1 - (-C_2^\epsilon)^{1/2} | \leq C\epsilon^2$$

and

$$| \epsilon^{-1} - x_N - (-C_2^\epsilon)^{1/2} | \leq C\epsilon^2,$$

respectively.

Proof. We first show that for any two consecutive zeros x, y of v_x,

$$\int_x^y \| v_x \mid -1 \mid dx \leq C\epsilon^2. \tag{46}$$

Assume for convenience that $v_x > 0$ on (x, y). By Lemma 3.4 it suffices to show that

$$\int_{y-2\tau}^{y} \mid v_x - 1 \mid dx \leq C\epsilon^2,$$

as the estimate over $(x, x + 2\tau)$ is similar (recall that $\tau = 2\epsilon^2 \ln 2\epsilon^{-1}$). Let z be the first point in $(y - 2\tau, y)$ such that $v_x = 1 - \overline{C}\epsilon$, where $\overline{C}$ is the constant appearing in Lemma 3.3. The integral over $(y - 2\tau, z)$ is readily estimated. Finally we define points $y_0 = z < y_1 < \ldots < y_p$ in the interval (z, y) by $v_x(y_k) = 1 - 2^k\overline{C}\epsilon$, where $1/2 \leq 2^p\overline{C}\epsilon < 1$. The points y_k are well defined by Lemma 3.3, and a 'time-map' estimate like (40) gives $\mid y_{k+1} - y_k \mid \leq C\epsilon^2$, from which the assertion follows after a short calculation.

To show (44) we will apply (38) at $x \pm 2\tau$. Assuming for convenience $v_x(x - 2\tau) > 0$, we have, by the second part of Lemma 3.4,

$$\begin{aligned} v_x(x \pm 2\tau) &= \mp 1 + \mathcal{O}(\epsilon^2), \\ v_{xx}(x \pm 2\tau) &= \mathcal{O}(1), \end{aligned}$$

and by (46)

$$\begin{aligned} v(x \pm 2\tau) &= v(x) - 2\tau + \mathcal{O}(\epsilon^2) \\ V(x \pm 2\tau) &= V(x) + \int_x^{x \pm 2\tau} v\, dx \\ &= V(x) \pm 2\tau v(x) + \mathcal{O}(\epsilon^2). \end{aligned}$$

Substituting this into (38), (44) follows.

Note that at $v_x = 0$, v has a (local) extremum. Therefore (44) in particular implies that $\max \mid v \mid \leq (-C_2^\epsilon + C\epsilon^2)^{1/2}$. Combining this with (46), it follows that the maximal distance of two consecutive zeros of v_x is bounded from above by $(-C_2^\epsilon + C\epsilon^2)^{1/2} + C\epsilon^2$. By Lemma 3.3 and Lemma 2.3 the energy estimate (18) can thus only hold if $-C_2^\epsilon$ is bounded from below by a positive constant, uniformly in ϵ. Using this fact in combination with (44) we see that $v(x) = \pm(-C_2^\epsilon)^{1/2} + \mathcal{O}(\epsilon^2)$ whenever $v_x = 0$. Now (45) follows from (46). The result concerning the first and the last zero of v_x follows from (46) as well, since $v(0) = v(\epsilon^{-1}) = 0$.

We are now ready to prove the main result.

Proof of Theorem 1.1. In view of the previous lemma and the scaling (12) only two things remain to be checked. First we have to ensure that the zeros of v_x are well separated, secondly we have to verify that $2(-C_2^\epsilon)^{1/2} = L_0 + \mathcal{O}(\epsilon)$. Intuitively it is obvious that for a *minimizing* v two zeros of v_x cannot be very close since such a pair of zeros would contribute an energy of $\sim 2A_0$ (see Lemmas 2.3 and 3.3). Removing the pair only effects a tiny change in v and should therefore lower the energy. Rather than giving all the technical details of the exact argument we illustrate it at a typical example. Let x_0, x_1, x_2, x_3 be consecutive zeros of v_x such that $x_1 - x_0 \geq c$, $x_3 - x_2 \geq c$, while $x_2 - x_1 \leq 4\tau = 8\epsilon^2 \ln 2\epsilon^{-1}$.

Let ϕ be a smooth function with support in $(x_0 + 2\tau, x_1 - 2\tau)$ such that $0 \leq \phi \leq 1$, $|\phi_x| \leq C, \int \phi \, dx \geq c/2$. let $y_1 = x_1 - 2\tau, y_2 = x_2 + 2\tau$ and let

$$w = \begin{cases} v_x(y_1) + \dfrac{v_x(y_2) - v_x(y_1)}{y_2 - y_1}(x - y_1) & \text{if } x \in (y_1, y_2), \\ v_x + s\phi & \text{else,} \end{cases}$$

where

$$s = \left(\int_{-\infty}^{\infty} \phi \, dx\right)^{-1} \int_{y_1}^{y_2} (v_x - w) \, dx = \mathcal{O}(\tau).$$

We have $w = v_x$ outside $(x_0 + 2\tau, y_2)$, and inside that interval $|w \pm 1| \leq C\tau$ and $|w_x| \leq C$, since $v_x(y_i) = \pm 1 + \mathcal{O}(\epsilon^2)$, by Lemma 3.4. Moreover $\int_{x_0}^{y_2}(v_x - w)\, dx = 0$, so that the function $\tilde{v}$ defined by

$$\tilde{v}(x) = v(x_0) + \int_{x_0}^{x} w \, dx,$$

coincides with v outside (x_0, y_2). A straightforward calculation shows that $\tilde{v}$ has lower energy than v.

We finally show $2(-C_2^\epsilon)^{1/2} = L_0 + \mathcal{O}(\epsilon)$. Let N be the number of zeros of v_x and let $d_1 = \epsilon^{-1}N^{-1}$. By Lemma 3.5, $d_1 = 2(-C_2^\epsilon)^{1/2} + \mathcal{O}(\epsilon^2)$. Let x_i, x_{i+1} be two consecutive zeros of v_x. Repeating the argument in the proof of Lemma 2.5 with the improved estimates (42) and (46) we obtain

$$J^\epsilon(v; x_i, x_{i+1}) \geq A_0 + \frac{d_1^3}{12} - C\epsilon^2.$$

A similar calculation (using $v(0) = v(\epsilon^{-1}) = 0$) shows that

$$J^\epsilon(v; 0, x_1) + J^\epsilon(v; x_N, \epsilon^{-1}) \geq A_0 + \frac{d_1^3}{12} - C\epsilon^2.$$

Therefore

$$\begin{aligned} J^\epsilon(v; 0, \epsilon^{-1}) &\geq N\left(A_0 + \frac{d_1^3}{12} - C\epsilon^2\right) \\ &= \epsilon^{-1}\left(A_0 d_1^{-1} + \frac{d_1^2}{12} - C\epsilon^2\right). \end{aligned}$$

Now $d \mapsto A_0 d^{-1} + d^2/12$ has a (quadratic) minimum at $d = L_0$ with minimum value E_0 and hence (25) implies that $|d_1 - L_0| \leq C\epsilon$ as claimed. This completes the proof of Theorem 1.1.

Acknowledgements

In the preparation of this work I profited from many helpful and encouraging discussions with John Ball, Jack Carr and Bob Pego. I would also like to thank the organizers of the GAMM International Conference on Problems Involving Change of Type for the invitation to present these results at the conference honouring Jack Hale. I gratefully acknowledge support through a grant from Cusanuswerk and additional support from the EEC programme on Mathematical Problems in Nonlinear Mechanics.

References

[BJ 87] J.M.Ball and R.D.James, Fine phase mixtures as minimizers of the energy, Arch. Rat. Mech. Anal. **100** (1987), 13–52.

[BJ 89] J.M.Ball and R.D.James, Crystal microstructure and the two-well problem, forthcoming.

[CGS 84] J.Carr, M.E.Gurtin and M.Slemrod, Structured phase transitions on a finite interval, Arch. Rat. Mech. Anal. **86** (1984), 317–351.

[CH 58] J.W.Cahn and J.E.Hilliard, Free energy of a nonuniform system. I. Interfacial free energy, J. Chem. Physics **28**, 258–267.

[KS 88] R.V.Kohn and P.Sternberg, Local minimizers and singular perturbations, to appear in: Proc. Roy. Soc. Edinburgh.

[Mo 87] L.Modica, Gradient theory of phase transitions and minimal interface criteria, Arch. Rat. Mech. Anal. **98** (1987), 357–383.

[Ta 87] L.Tartar, unpublished note.

[vdW 93] J.D. van der Waals, The thermodynamic theory of capillarity under the hypothesis of a continuous variation in density (in Dutch), Verhandel. Konink. Akad. Weten. Amsterdam (Sec. 1) Vol. 1(1893), No. 8.

2. Phasetransition and Dynamics

DYNAMICS OF CLUSTER GROWTH

J. Carr
Department of Mathematics
Heriot-Watt University
Edinburgh EH14 4AS
Scotland, U.K.

§1. INTRODUCTION

The formation of a distribution of cluster sizes is a common feature in a wide variety of systems. Examples include astrophysics, atmospheric physics, colloidal chemistry, polymer science and the kinetics of phase transitions in binary alloys. In this paper we discuss the mathematical theory of a model for the dynamics of cluster growth. The processes described by this model involve coagulation of clusters via binary interactions and fragmentation, a unimolecular process. The distribution of cluster sizes is determined by the competition between these processes.

If $c_j(t) \geq 0$, $j=1,2,\ldots,$ denotes the expected number of j-particles per unit volume, then the discrete coagulation-fragmentation equations are

$$\dot{c}_j = \frac{1}{2}\sum_{k=1}^{j-1} [a_{j-k,k}\, c_{j-k}\, c_k - b_{j-k,k}\, c_j] - \sum_{k=1}^{\infty} [a_{j,k}\, c_j\, c_k - b_{j,k}\, c_{j+k}] \qquad (1.1)$$

for $j=1,2,\ldots$ The coagulation rates $a_{j,k}$ and fragmentation rates $b_{j,k}$ are nonnegative constants with $a_{j,k} = a_{k,j}$ and $b_{j,k} = b_{k,j}$. In equation (1.1) the first two terms represent the rate of change of the j-cluster due to the coalescence of smaller clusters and the breakup of the j-cluster into smaller clusters. The final two terms represent the change due to coalescence of the j-cluster with other clusters and the breakup of larger clusters into j-clusters. For a derivation of this equation and its analogue in which the cluster size is a continuous variable see [6]. The model neglects the geometrical location of clusters and spatial fluctuations in cluster density. For further information on these effects see [4, 5].

§2. SOME SPECIAL CASES

Since matter is neither destroyed nor created in the interactions described by (1.1) we expect that the density $\rho = \sum j\, c_j(t)$ is a conserved quantity. In certain circumstances however, the density is not conserved. To illustrate this and other phenomena we consider some special cases.

(a) Let $b_{j,k} = 0$ for all j and k so that we only consider coagulation. We further specialise to two forms of coagulation kernel:

$$a_{j,k} = j^{\alpha} + k^{\alpha} \tag{2.1}$$

$$a_{j,k} = (jk)^{\alpha} \tag{2.2}$$

The additive form of $a_{j,k}$ in (2.1) would arise in applications if we assumed that binary interactions occur randomly with a rate depending on the effective surface area. The multiplicative form of (2.2) would apply to situations in which bond linking was the dominant mechanism. Note that for the kernel (2.1), large-large and large-small interactions have the same order of magnitude (i.e. $a_{j,k} \cong a_{j,j}$ for large j and small k), whereas for (2.2) large-large interactions dominate.

If $\alpha > 1/2$, then for the kernel (2.2), density conservation can break down in finite time [7]. This is interpreted as the appearance of an infinite cluster or gel. For the kernel (2.1), if a solution exists then density is conserved [3].

To gain some insight into the dependence of the rate of growth of clusters we use a technique due to Leyvraz and Tschudi [9] to relate solutions of (1.1) with different initial data. We first consider the kernel (2.1) so that (1.1) takes the form

$$\dot{c}_j = \frac{1}{2}\sum_{k=1}^{j-1} [(j-k)^{\alpha} + k^{\alpha}]\, c_{j-k}\, c_k - \sum_{k=1}^{\infty} (j^{\alpha} + k^{\alpha})\, c_j\, c_k \tag{2.3}$$

Let c_j^1 be a solution of (2.3) with initial data $c_j^1(0) = \delta_{j,1}$. For positive integers n define $c^n(t) = (c_j^n(t))$, j=1,2,... by

$$\begin{aligned} c_{nj}^n(t) &= n^{-1} c_j^1(n^{\alpha-1}t) \\ c_r^n(t) &= 0, \quad r \text{ not a multiple of } n. \end{aligned} \tag{2.4}$$

It is then easy to check that $c^n(t)$ is a solution of (2.3) with initial data given by $c_j^n(0) = n^{-1}\delta_{j,n}$. From (2.4) we see that the time scale for this class of solutions depends on the sign of $\alpha - 1$. In fact, if $\alpha \le 1$, we get global existence for the general initial value problem (with initial data having finite density) and nonexistence of global solutions if $\alpha > 1$ [3].

For the kernel (2.2), let c^1 be the solution of (1.1) with initial data $c_j^1(0) = \delta_{j,1}$. It is shown in [9] that the appropriate scaling is

$$c_{nj}^n(t) = n^{-1} c_j^1(n^{2\alpha-1} t), \qquad c_r^n(t) = 0 \text{ otherwise.} \tag{2.5}$$

From (2.5), we see that $\alpha = 1/2$ is the critical parameter. Global solutions for the initial value problem exist for $\alpha \leq 1$ (see [8] for a proof), but as remarked above, density conservation breaks down after a finite time if $\alpha > 1/2$. It is interesting to note that if $\alpha > 1$ we can still have global existence for this case.

(b) Let $a_{j,k} = 0$ for all j,k so that we consider the linear fragmentation equations. For any initial data with finite density, (1.1) has a density conserving solution. However, for a large class of fragmentation coefficients (for example $b_{j,k} = (j+k)^\beta$, $\beta > -1$), equation (1.1) has solutions with density $e^{t\lambda}$ for any $\lambda > 0$. In particular, solutions need not be unique. These spurious solutions are not of physical interest and this leads to the mathematical problem of finding a criterion for selecting the correct solution for the general equation (1.1).

(c) Consider the Becker-Döring equation for which $a_{j,k} = b_{j,k} = 0$ if both j and k are greater than 1. The mathematical theory of these equations has been studied in [1]. In this case the density is always a conserved quantity. The asymptotic behaviour of solutions is interesting both mathematically and for applications. Under certain hypotheses on the rate coefficients and the density of the initial data ρ_0, we have that

$$\rho_0 = \sum_{j=1}^{\infty} j\, c_j(t) > \sum_{j=1}^{\infty} j \lim_{t\to\infty} c_j(t) = \rho_s .$$

The excess density $\rho - \rho_s$ corresponds to the formation of larger and larger clusters as $t \to \infty$ and may be interpreted as a transition from microscopic to macroscopic clusters. Mathematically this can be identified with a weak but not strong convergence as $t \to \infty$. See also [10] for an analysis of metastable solutions and [2, 11] for some technical refinements.

§3. RESULTS

In the previous section, some of the analytical difficulties associated with (1.1) were discussed. In order to generalise the results for the Becker-Döring Equation to the more general equation, we have to restrict attention to a subclass of coagulation-fragmentation kernels. In this section we outline some recent results concerned with existence and density conservation [3]. We first introduce some notation. Let

$$X^{+} = \{\, y = (y_r) \;:\; \|y\| < \infty,\ \text{each } y_r \geq 0 \,\}, \qquad \|y\| = \sum_{r=1}^{\infty} r\,|c_r|.$$

<u>THEOREM</u> <u>1</u> Assume that $a_{j,k} \leq K(j + k)$, for all j,k where K is a constant. Let $c_0 \in X^{+}$. Then there exists a solution c of (1.1) on $[0,\infty)$ with $c(0) = c_0$.

We prove the above result by taking a limit of solutions of the finite dimensional system

$$\dot{c}_j = \frac{1}{2}\sum_{k=1}^{j-1} [a_{j-k,k}\, c_{j-k}\, c_k - b_{j-k,k}\, c_j] - \sum_{k=1}^{n-j} [a_{j,k}\, c_j\, c_k - b_{j,k}\, c_{j+k}] \tag{3.1}$$

where $1 \leq j \leq n$.

In general, even if $a_{j,k} \leq K(j + k)$, solutions of (1.1) are not unique and do not conserve density. Thus a criterion is needed to single out physically meaningful solutions. We call a solution to (1.1) <u>admissible</u> if it is the limit of solutions to the truncated system (1.1). The next result shows that for $a_{j,k} \leq K(j + k)$, this criterion excludes the non-physical solutions described in the previous section in that any admissible solution conserves density.

<u>THEOREM</u> <u>2</u> Assume that $a_{j,k} \leq K(j + k)$ and that c is an admissible solution of (1.1) on $[0,T)$. Then for all $t \in [0,T)$,

$$\sum_{j=1}^{\infty} j\, c_j(t) = \sum_{j=1}^{\infty} j\, c_j(0).$$

It is also useful to have conditions under which all solutions of (1.1) conserve density.

<u>THEOREM</u> <u>3</u> Assume that for some $n_0 \geq 0$ and $K \geq 0$ we have

(*i*) $a_{j,k} = r_j + r_k + \alpha_{j,k}$ where $\{r_j\}$ is a nonnegative sequence and

$\alpha_{j,k} \le K(j + k)$ for all $j,k \ge n_0$.

(*ii*) $\sum_{j=1}^{p} j\, b_{r-j,j} \le K\, r, \quad r \ge 2\, n_0$ where p is the integer part of $(r + 1)/2$.

Then if c is a solution of (1.1) on $[0,T)$ with $\rho_0 = \sum_{j=1}^{\infty} j\, c_j(0) < \infty$,

$$\sum_{j=1}^{\infty} j\, c_j(t) = \rho_0 \quad \text{for all } t \in [0,T).$$

The hypothesis (*ii*) has an interesting physical interpretation. If surface energy effects are important then it is unlikely that a large cluster of size $j + k$ will break up into two large clusters as this would increase the surface energy by a large amount. Thus $b_{j,k}$ should be small if j and k are both large. Condition (*ii*) is a precise formulation of these heuristics.

REFERENCES

1. Ball, J.M., Carr, J., Penrose, O.: The Becker-Döring cluster equations: basic properties and asymptotic behaviour of solutions. Commun. Math. Phys. **104**, 657-692 (1986).

2. Ball, J.M., Carr, J.: Asymptotic behaviour of solutions to the Becker-Döring equations for arbitrary initial data, Proc. Roy. Soc. Edinburgh 108A, 109-116 (1988).

3. Ball, J.M., Carr, J.: In preparation.

4. Binder, K. : Theory for the dynamics of clusters. II. Critical diffusion in binary systems and the kinetics of phase separation. Phys. Rev. B 15, 4425-4447 (1977).

5. van Dongen, P.G.J.: Spatial fluctuations in reaction-limited aggregation. J. Stat. Phy. 54, 221-271 (1989).

6. Drake, R.: In: Topics in current aerosol research. International reviews in aerosol physics and chemistry, Vol. 2, Hidy, G.M., Brock J.R. (eds.). Oxford: Pergamon Press (1972).

7. Hendricks, E.M., Ernst, M.H, Ziff, R.M.: Coagulation equations with gelation, J. Stat. Phys. 31, 519-563 (1983).

8. Leyvraz, F., Tschudi, H.R.: Singularities in the kinetics of coagulation processes. J. Phys. A:Math Gen. 14, 3389-3405 (1981).

9. Leyvraz, F., Tschudi, H.R.: Critical kinetics near gelation. J. Phys. A:Math Gen. 15, 1951-1964 (1982).

10. Penrose, O.: Metastable states for the Becker-Döring cluster equations, preprint.

11. Slemrod, M.: Trend to equilibrium in the Becker-Döring cluster equations, preprint.

A Geometric Approach to the Dynamics of $u_t = \varepsilon^2 u_{xx} + f(u)$ for small ε

G. FUSCO*

Dipartimento di Metodi Matematici
Via A. Scarpa 10 00161 Roma, ITALY
and
Department of Mathematics
University of Tennessee
Knoxville, TN 37916, USA

1. Introduction.

In some important cases the differential equations describing the evolution of a physical system contain a small parameter $\varepsilon > 0$ and the equations depend on ε in such a singular way as $\varepsilon \to 0$, that it is impossible to define limit differential equations which capture the global limit behavior of the system. In fact as $\varepsilon \to 0$ the dynamics of the system becomes more and more complex so that there is no limit dynamics and therefore no limiting differential equations. Examples of this situation are the Navier-Stokes equations (with ε the inverse of the Reynold's number) and certain differential equations from the theory of phase transitions. In dealing with problems of such a singular nature, one can still try to define some limit dynamical behavior for $\varepsilon \to 0$, provided attention is restricted to special features of the dynamics which happen to possess some kind of continuous dependence in ε as $\varepsilon \to 0$. The aim of this work is to show how this can be done effectively in the case of the scalar parabolic equation

$$(1) \qquad \begin{cases} u_t = & \varepsilon^2 u_{xx} + f(u), \quad x \epsilon (0,1), \\ & u_x = 0, \quad x = 0, 1 \ . \end{cases}$$

*Supported in part by The Science Alliance, a Program of the Tennessee Centers of Excellence and The National Science Foundation.

This equation is the gradient system corresponding to the Liapounov functional

$$J(u) = \int_0^1 \left(\frac{\varepsilon^2}{2} u_x^2 + F(u) \right) dx ,$$

with $f = -F'$, and it is perhaps the simplest mathematical model for the coexistence of two phases of the same substance at the transition temperature. In this context u is an "order parameter" which is related to the microscopic structure of the matter in such a way that u near -1 corresponds to one of the two phases (solid) and u near 1 corresponds to the other phase (liquid). $F(u)$ is the specific free energy of the matter and is assumed to have two equal minima at $u = \pm 1$. This corresponds to the fact that by definition at the transition temperature the two phases have the same free energy. The term $\varepsilon^2 u_x^2/2$ in the expression of $J(u)$ is added to penalize high gradients of the order parameter and therefore to model the tendency of the substance to minimize the number of interfaces separating the two phases. More information about equation (1) and more sophisticated models for phase transitions can be found in **[CF]** **[FG]** **[G1]** **[G2]**.

Equation (1) is an example of the singular situation described above. It is well known that it generates a dissipative semiflow in several function spaces (for instance in $W^{1,2}$) and that it possesses a global attractor $\mathcal{A}_\varepsilon$ **[He]** **[H]**. There is no global limiting system as $\varepsilon \to 0$, and in fact $\mathcal{A}_\varepsilon$ goes through a cascade of bifurcations with its structure becoming increasingly complex. For instance the number of fixed points (stationary solution of (1)) increases without bound as $\varepsilon \to 0$ and the same is true for the dimension of $\mathcal{A}_\varepsilon$ (**[He]** **[H]**). Therefore instead of considering the whole attractor, we fix our attention to the unstable manifolds of equilibria and try to analyze what happens to these sets and to the dynamics on them as $\varepsilon \to 0$. We show , Theorems 1 and 2 below, that, for $\varepsilon \ll 1$, a large part, $\widehat{W}_N$, of the unstable mainfold $W^u(\phi_N)$ of an equilibrium ϕ_N of (1) with N zeros is made of functions which are approximately step functions with values ± 1 and exactly N transition layers where sharp jumps occur. We also show that the dynamics on $\widehat{W}_N$ is described by a system of ODEs which also determines the motion of these interior layers. Since this motion is very slow for $\varepsilon \ll 1$ (speed of the order $O\left(e^{-\frac{c}{\varepsilon}}\right)$ for some $c > 0$) $\widehat{W}_N$ is called a "slow motion mainfold" SMM. The equations describing the dynamics of interior layers were derived in [**FH**] by solving an equation for the approximate description of SMM. We refer to [**FH**] also for a general discussion of the global dynamics of (1) for $\varepsilon \ll 1$. The derivation of the equations of motion of the transition layers has been established in [**CP**]. By deriving appropriate estimates these authors are able to construct, a locally positively invariant set which provides a good approximation to our invariant set $\widehat{W}_N$.

The technique we use in the proof of theorem 1 is based on geometric ideas contained in [**FH**]. See also [**F**] for some related work on the dynamical equivalence between scalar parabolic equations and certain systems of ODEs.

It turns out that the mainsteps in the proof of theorem 1 have an abstract character that allows a variety of applications as for instance proving the existence of a slow motion manifold for the Cahn-Hilliard equation [**ABF**].

2. Notation and main results.

To state our results on the asymptotic behavior as $\varepsilon \to 0$, of unstable mainfolds of equilibria of equation (1) we need some defintions and some notation.

As stated above we assume that F has two equal minima at $u = \pm 1$. We also assume that F is continuous and that $f = -F'$ has only three nondegenerate zeroes. For simplicity we shall work under the hypothesis that f is an odd function. The assumption that F has two equal minima at $u = \pm 1$ implies the existence of a unique solution U to the problem

$$(2) \qquad \begin{cases} \varepsilon^2 U_{xx} + f(U) = 0, & x \epsilon \boldsymbol{R} \\ U(0) = 0; & \lim_{x \to \pm\infty} U = \pm 1, \end{cases}$$

This solution corresponds to the existence of a heteroclinic connection between $(-1,0)$ and (1,0) in phase space for equation (2). Clearly U depends on ε and it is in fact a function of x/ε. To keep the notation simple we don't make explicit this dependence of the function u and of many other functions considered in the rest of the paper. For any small number $\rho, \rho > 0$ let $\Gamma\rho \subset \boldsymbol{R}^N$ be the set of vectors $\xi = (\xi_1, \dots, \xi_N)$ defined by $\Gamma\rho = \{\xi | 0 < \xi_1 < \dots < \xi_N < 1, \xi_{i+1} - \xi_i > 2\rho, i = 0, \dots, N\}$, with $\xi_0 = -\xi_1, \xi_{N+1} = 2 - \xi_N$. Let $\eta_i = (\xi_{i+1} + \xi_i)/2 \; i = 0, \dots, N$. For each ξ in $\Gamma\rho$ define $u^\xi : [0,1] \to \boldsymbol{R}$ by

$$(3) \qquad u^\xi(x) = (-1)^{i+1} U(x - \xi_i), \quad x \epsilon [\eta_{i-1}, \eta_i], \; i = 1, \dots, N,$$

u^ξ is a continuous function with a piecewise continuous first derivative. The map $\xi \to u^\xi$ defines a N-dimensional imbedded manifold $\overline{W}$ in $L^2 = L^2(0,1)$. If f is of class C^2 then $\overline{W}$ is a $C^{1,1}$ manifold. We have in fact the following Lemma

LEMMA 1. *Let a small number $\rho > 0$ and an integer $N \geq 1$ be fixed and assume f is C^2. Then*

$$C\varepsilon^{+\frac{1}{2}} \leq \|u_i^\xi\|_{L^2} \leq C^1 \; \varepsilon^{+\frac{1}{2}}, \quad j = 0, 1$$

$$\|u_i^{\bar{\xi}} - u_i^\xi\| \leq C\varepsilon^{-2} |\bar{\xi} - \xi|$$

for some constant C, C^1 independent of $\varepsilon, \xi \epsilon \Gamma\rho, u_i^\xi := \frac{\partial u^\xi}{\partial \xi_i}$.

PROOF: By the definition of u^ξ it follows that $u_i^\xi(x) = (-1)^i U_x(x - \xi_i), x \epsilon (\eta_{i-1}, \eta_i); u_i^\xi(x) = 0, x \notin [\eta_{i-1}, \eta_i]$. From this and simple estimates on U the lemma follows. ∎

Let $\gamma = (\gamma_1, \ldots, \gamma_N) : \Gamma\rho \to \boldsymbol{R}^N$ be the function defined by

$$\gamma_i(\xi) = \varepsilon K \left[\exp\left(-\frac{\nu}{\varepsilon}(\xi_{i+1} - \xi_i)\right) - \exp\left(-\frac{\nu}{\varepsilon}(\xi_i - \xi_{i-1})\right)\right], \quad i = 1, \ldots, N . \tag{4}$$

where $\nu^2 = -f'(\pm 1)$ and K is defined in terms of the function

$$g(u) = 2(F(u) - F(-1)) = -2 \int_{-1}^{u} f \tag{5}$$

by

$$K = \frac{\nu^2 \exp\left[2\nu \int_0^1 \left(g^{-\frac{1}{2}} - \frac{1}{\nu(1-u)}\right)\right]}{\int_0^1 g^{\frac{1}{2}}}$$

In the special case $f(u) = u(1 - u^2)$ it $\nu = \sqrt{2}, K = \frac{3\sqrt{2}}{4}$. We can now state our main results.

THEOREM 1. *Let a small real $\rho > 0$ and an integer $N \geq 1$ be fixed. Then for each $\varepsilon > 0$ sufficiently small there exist functions*

$$\Gamma\rho \ni \xi \to \hat{u}^{\xi} \epsilon\, W^{1,2},$$

$$\Gamma\rho \ni \xi \to \hat{\gamma}(\xi) \epsilon\, \boldsymbol{R}^n,$$

with the following properties

(i) $\xi \to \hat{u}^{\xi}, \xi \to \hat{\gamma}_{\xi}(\xi)$ *are Lipschitz continuous.*

(ii) $\hat{W} = \{u | u = \hat{u}^{\xi}, \xi \epsilon \Gamma\rho\}$ *is an invariant manifold for the dynamical system defined by (1) in* $W^{1,2}$.

(iii) *The flow on* $\widehat{W}$ *is described by the ordinary differential equation,*

$$\dot{\xi} = \hat{\gamma}(\xi)$$

(iv) *the following estimates hold*

$$\|\hat{u}^{\xi} - u^{\xi}\|_{W^{1,2}} = 0(\exp(-\frac{\nu}{\varepsilon}\delta^{\xi})) ,$$

$$\|\hat{\gamma}(\xi) - \gamma(\xi)\| = 0(\exp(-\frac{3\nu}{\varepsilon}\delta^{\xi}))$$

where δ^{ξ} is any number less that $\frac{1}{2}\min(\xi_i - \xi_{i-1})$.

THEOREM 2. *The invariant manifold $\hat{W} = \{u | u = \hat{u}^\xi, \xi \epsilon \Gamma \rho\}$ defined in theorem 1 contains a unique stationary solution $u^{\bar{\xi}}$ of (1), $u^{\bar{\xi}}$ is hyperbolic and $\hat{W}$ is an open subset of $W^u(u^{\bar{\xi}})$.*

In the remaining part of this paper we describe the proof of Theorem 1. Theorem 2 is essentially a consequence of Theorem 1.

PROOF OF THEOREM 1

In this section we present a proof of Theorem 1. Our intention here is to stress the systematic character of the proof by presenting only the main ideas and by discussing the simplest situation. Therefore in the proof we shall restrict ourselves to the case $N = 1$. A fully detailed proof for the general case $N \geq 1$ will appear elsewhere.

Step 1: Construction of a reference problem.

We aim to construct a reference problem (1)* such that, as far as existence of slow motion mainfolds is concerned, equation (1) can be considered a regular perturbation of (1)*. As we shall see, this can be done because, unlike sets containing single equilibria or the whole attractor, SMM's are indeed the right objects to be studied as $\varepsilon \to 0$. In fact they contain all directions of asymptotically vanishing eigenvalues and therefore behave as "normally hyperbolic sets", uniformly in $\varepsilon \to 0$. Lemma 2 below makes this statement more precise.

A natural choice of the reference problem is the following.

$$(1^*) \qquad \begin{cases} u_t = \varepsilon^2 u_{xx} + f(u), & x \epsilon (0,1) \\ u_x(x) = \phi_0(u), & u_x(1) = \phi_1(u)\ , \end{cases}$$

where $\phi_0, \phi_1 : W^{2,2} \to \boldsymbol{R}$, are smooth nonlinear functionals satisfying the conditions

$$(6) \qquad \phi_0(u^\xi) = u_x^\xi(0); \; \phi_1(u^\xi) = u_x^\xi(1)\ .$$

From the definition of u^ξ and u it follows that $u^\xi_x(0)$ and $u^\xi_x(1)$ are $0(\exp(-\frac{\nu}{\varepsilon}\rho))$. This implies that, as can be easily shown, we can choose ϕ_0, ϕ_1 so that they are globally bounded by $0(\exp(-\frac{\nu}{\varepsilon}\rho))$. Therefore we can expect that, for $\varepsilon \ll 1$, the dynamic of (1) should be close in some sense to the dynamics of $(1)^*$. On the other hand it is clear that by (6), and the definition of u^ξ, $\overline{W}$ is an invariant manifold for $(1)^*$. Actually $\overline{W}$ is a manifold of stationary solutions of $(1)^*$. It is therefore to be expected that for $\varepsilon \ll 1$ equation (1) has an invariant manifold $\hat{W}$ near $\overline{W}$ on which the motion should be very slow.

We remark that the choice of the reference problem can be made in many different ways. To give an example, one could define $\tilde{u}^\xi = \chi u^\xi$ with $x : [0,1] \to [0,1]$ a C^∞ function chosen so that $\tilde{u}^\xi$ satisfies the boundary conditions. Then the natural reference problem should be

$$(1^{**}) \qquad \begin{cases} u_t = \varepsilon^2 u_{xx} + f(u) + \phi(u), \\ u_x = 0, x = 0, 1, \end{cases}$$

where ϕ is a nonlinear functional satisfying the condition

$$\phi(\tilde{u}^\xi) = -(\varepsilon^2 \tilde{u}^\xi_{xx} + f(\tilde{u}^\xi)) \, .$$

This condition would in fact inply $\widetilde{W} = \{u | u = \tilde{u}^\xi\}$ is a manifold of equilibria for $(1)^{**}$ and in view of the fact that $\|\phi(\tilde{u}^\xi)\|_{L^2} = 0(\exp(-\frac{\nu}{\varepsilon}\rho))$ one should expect a slow motion manifold $\hat{W}$ near $\widetilde{W}$.

Step 2: Construction of a first approximation for $\hat{W}$ and for the flow on it.

On the basis of the discussion in step (1) it is natural to look for an invariant manifold $\hat{W}$ of equation (1) which is a graph over $\overline{W}$. Therefore we construct a tubular neighborhood $\overline{\mathcal{N}}$ of $\overline{W}$ by setting

$$(7) \qquad \begin{cases} u = u^\xi + v, \\ \left\langle v, u^\xi_\xi \right\rangle = 0, \end{cases}$$

where $\langle\cdot,\cdot\rangle$ is the standard inner product in L^2 and the subscript ξ denote differentiation with respect to ξ. A standard application of the implicit function theorem shows that (7) define a smooth change of variables from an open neighborhood $\overline{\mathcal{N}}$ of $\overline{W}$ onto an open neighborhood of the zero section $v = 0$ in the linear fibration $\mathcal{F} = \{\overline{F}^{\xi}\}, =$ $\{u | u\epsilon L^2(0,1), \left\langle u, u^{\xi}_{\xi}\right\rangle = 0\}$. It can also be shown that the size of $\overline{\mathcal{N}}$ and of the corresponding neighborhood can be taken independent of ε [**CP**].

A function $\xi \to \hat{u}^{\xi}\epsilon C^2[0,1]$ defines an invariant manifold $\hat{W} \subset L^2$ for (1) which is C^1 close to $\overline{W}$ if and only if there is a function $\xi \to \hat{v}^{\xi}$ which is C^1 close to the zero function and satisfies

$$(8) \qquad \begin{cases} \left(u^{\xi}_{\xi} + \hat{v}^{\xi}_{\xi}\right)\hat{\gamma}(\xi) = \varepsilon^2(u^{\xi}_{xx} + \hat{v}^{\xi}_{xx}) + f(u^{\xi} + \hat{v}^{\xi}) \\ \hat{v}^{\xi}_{x} = -u^{\xi}_{x}, \ x = 0,1 \\ \left\langle \hat{v}^{\xi}, u^{\xi}_{\xi}\right\rangle = 0 \end{cases}$$

for some C^1 function $\xi \to \hat{\gamma}(\xi)$.

We expect $\hat{v}^{\xi}$, $\hat{v}^{\xi}_{\xi}, \hat{\gamma}(\xi)$ to be very small quantities for $\varepsilon \ll 1$. Therefore a good approximation to $\hat{v}^{\xi}, \hat{\gamma}(\xi)$ should be obtained by solving the linear problem that one gets by retaining in equation (8) only linear terms in $(\hat{v}^{\xi}, \hat{v}^{\xi}_{\xi}, \hat{\gamma}(\xi))$. Since by definition of u^{ξ} we have $\varepsilon^2 u^{\xi}_{xx} = -f(u^{\xi})$, this linear problem is

$$(9) \qquad \begin{cases} u^{\xi}_{\xi}c(\xi) = \varepsilon^2 v^{\xi}_{xx} + f'(u^{\xi})v^{\xi} \\ v^{\xi}_{x} = -u^{\xi}_{x}, \ x = 0,1 \\ \left\langle v^{\xi}, u^{\xi}_{\xi}\right\rangle = 0 \end{cases}$$

where we have used the notation c, v^{ξ} instead of $\hat{\gamma}, \hat{v}^{\xi}$ to remark that solutions to equation (9) are only approximations to solutions of equation (8).

Proposition 1. [**FH**] *Assume $f\epsilon C^3$ and let a small number $\rho > 0$ be fixed. Then there is an $\overline{\varepsilon} > 0$ independent of $\xi\epsilon\Gamma\rho$ such that for any $\varepsilon < \overline{\varepsilon}$ and any $\xi\epsilon\Gamma\rho$ equation 9) has a* *unique solution $c(\xi), v^\xi$. The function $p^\xi = u^\xi + v^\xi$ is a C^4 function. The function $c \to c(\xi)$ is C^2, the function $\xi \to p^\xi$ is a C^2 function as a function from $\Gamma\rho$ to $W^{2,2}$. Moreover the following estimate holds*

$$c(\xi) = \gamma(\xi)(1 + 0(\exp\left(-\frac{\nu}{\varepsilon}\delta^\xi\right)), \quad \xi\epsilon\Gamma\rho$$

where γ is the function defined by 4) for $N = 1$ and δ^ξ any number $< \min(\xi, (1-\xi))$.

The following proposition is a precise statement substantiating the conjecture that the set $\widehat{W}$ should behave as a normally hyperbolic set even in the limit for $\varepsilon \to 0$. For $\xi\epsilon\Gamma\rho$ we let $\overline{P}^\xi, \overline{Q}^\xi$ be the projection of L^2 defined by

$$\overline{P}^\xi\phi = \frac{\left\langle \phi, u_\xi^\xi \right\rangle}{\left\langle u_\xi^\xi, u_\xi^\xi \right\rangle} u_\xi^\xi, \qquad \overline{Q}^\xi = I - \overline{P}^\xi . \tag{10}$$

Proposition 2. *Let a small $\rho > 0$ be fixed. For each $\xi\epsilon\Gamma\rho$ and for each $\phi\epsilon C^2[0,1] \cap \overline{F}^\xi$, $\phi_x = 0, x = 0,1$, let $\overline{A}^\xi\phi$ be defined by*

$$\overline{A}^\xi\phi = \overline{Q}^\xi\ (\varepsilon^2\phi_{xx} + f'(u^\xi)\phi) . \tag{11}$$

Then $\overline{A}^\xi$ can be extended to a selfadjoint operator of $\overline{F}^\xi = \overline{Q}^\xi L^2(0,1)$ and there is a constant $C > 0$ independent of ε and ξ such that

$$spectr\ \overline{A}^\xi \leq -C \tag{12}$$

PROOF: For a proof, based in part on arguments of [**NF**], see [**ABF**]. A different proof using the angle argument is given in [**CP**].

Lemma 2. *Assume $f\epsilon C^3$. Then, if $\varepsilon > 0$ is sufficiently small, the following estimates hold*

$$(i) \qquad \|\frac{d^j}{d\xi^j}\ v^\xi\|_{W^{2,2}} < C\ \exp(-\frac{\nu}{\varepsilon}\delta^\xi), \quad j = 0,1,2$$

$$(ii) \qquad |v^\xi(x)| < C\ \exp(-\frac{3\nu}{2\varepsilon}\delta^\xi), x\epsilon(\xi - \varepsilon^{\frac{1}{2}}, \xi + \varepsilon^{\frac{1}{2}})$$

for some constant $C > 0$ independent of ε, ξ.

PROOF: Case $j = 0$ of (i) follows in a standard way from equation (9) and Proposition 2. The case $j = 1, 2$ is then obtained inductively by differentiating (9). The inequality (ii) is proved by analyzing the explicit expression of v^ξ given in [FH].

Before proceeding to the next point we make some remarks on the meaning of the approximation to the manifold $\widehat{W}$ constructed above. In fact we can ask how good an approximation should be in order to establish existence of $\widehat{W}$ by means of invariant manifold theory. We believe that the degree of accuracy of the approximation is dectated by the accuracy with which we wish to describe the vector field on $\widehat{W}$. If we merely want to show that this vector field is $0(\exp -\frac{c}{\varepsilon})$ the first approximation to $\widehat{W}$ can be identified with the mainfold $\overline{W}$ and the vector field $\hat{\gamma}$ on $\widehat{W}$ can be approximated with the zero vector field. If instead one is willing to show that $\hat{\gamma} = \gamma + e$ with e satisfying an estimate as in Theorem 1, then the first approximation must be chosen more carefully as we have done above by solving equation (9). If one wants to move a step further in the asymptotic expansion of e then the first approximation constructed above will not be sufficiently accurate and a better approximation is needed. A sequence of successive approximations $(c^k, u^{\xi,k})$, $k = 0, 1,$ can be constructed by setting

$$c^0 = c; \; w^{\xi,0} = 0$$

$$u^{\xi,k} = u^\xi + v^\xi + w^{\xi,k}, \quad k = 1, \ldots$$

and by computing $c^k, w^{\xi k}$ by solving the problem

$$\begin{cases} u^\xi_\xi c^k = \varepsilon^2 w^{\xi,k}_{xx} + f'(u) w^{\xi,k} + r\left(\xi, c^{k-1}, w^{\xi,k-1}, w^{\xi,k-1}_\xi\right) \\ w^{\xi,k}_x = 0, \quad x = 0, 1, \quad \left\langle w^{\xi,k}, u^\xi_\xi \right\rangle = 0 \end{cases}$$

where

$$r(\xi, a, y, z) = f(u^\xi + v^\xi + y) - f(u^\xi) - f'(u^\xi)(v^\xi + y) + a(u^\xi_\xi + v^\xi_\xi + z)$$

Notice that Proposition 2 above implies this problem can be solved and that

$$w^{\xi,k} = (\overline{A}^\xi)^{-1}\overline{Q}^\xi\, r(\xi, c^{k-1},\ w^{\xi,k-1}, w_\xi^{\xi,k-1})\,,$$

$$c^k u^\xi_\xi = \overline{P}^\xi(\epsilon^2 w^{\xi,k}_{xx} + f'(u)w^{\xi,k} + r(\xi, c^{k-1}, w^{\xi,k-1}, w_\xi^{\xi,k-1}))\,.$$

As a final remark we observe that one could try to use the above procedure for computing $(c^k, u^{\xi,k})$ and obtain the solution $(\hat{\gamma}, \hat{u}^\xi)$ as the limit of $(c^k, u^{\xi,k})$. This approach cannot be successful without some adjustment since it entails differentiation of w with respect to ξ and therefore a loss of one derivative at each step. It is likely that some modification of the simple iteration scheme above and an application of the hard implicit function theorem can resolve this difficulty. Here we follow a different approach based on invariant manifold theory.

Step 3: Construction of an invariant manifold $\widehat{W}$ near $W = \{u | u = p^\xi = u^\xi + v^\xi, \xi \epsilon \Gamma p\}$.

Standard methods of invariant manifolds theory do not seem to be directly applicable to the present situation. Therefore, several adjustments of the standard approach are needed which complicate the construction considerable. As a result we develop a technique which can be used also in other similar problems in singular perturbation theory. We describe this technique by subdividing Step 3 in several points.

POINT 1: Introduce new coordinates (ξ, w) in a tubular neighborhood $\mathcal{N}$ of W and derive from equations (1) differential equations for ξ, w.

As was done before with respect to $\overline{W}$ we can define a tubular neighborhood $\mathcal{N}$ of W by setting

$$\left\{\begin{array}{l} u = p^\xi + w, \\ \left\langle w, p^\xi_\xi \right\rangle = 0 \, . \end{array}\right. \tag{13}$$

This equations define a smooth change of variables from $\mathcal{N}$ onto an open neighborhood of the zero section $w = 0$ in the linear fibration

$$\mathcal{F} = \{F^\xi, \xi\epsilon\Gamma\rho\}, \; F^\xi = \{w/w\epsilon L^2(0,1), \left\langle w, p^\xi_\xi \right\rangle = 0\} \, .$$

For $\xi\epsilon\Gamma\rho$ let P^ξ, Q^ξ be the projection on L^2 defined as in (10) with u^ξ_ξ replaced by p^ξ_ξ. By letting $u = p^\xi + w$ in equation (1) we get

$$p^\xi_\xi \, \dot{\xi} + w_t = \varepsilon^2 w_{xx} + \varepsilon^2 p^\xi_{xx} + f(p^\xi + w) \, . \tag{14}$$

By definition of p^ξ and equation (9) it follows

$$\varepsilon^2 p^\xi_{xx} = u^\xi_\xi c(\xi) - f(u^\xi) - f'(u^\xi) v^\xi \, . \tag{15}$$

Therefore, if we take the scalar product of equation (14) with p^ξ_ξ, and solve with respect to $\dot{\xi}$, and taking account that $\left\langle w, p^\xi_\xi \right\rangle = 0$ implies

$$\left\langle w_t, p^\xi_\xi \right\rangle = -\dot{\xi} \left\langle w, p^\xi_{\xi\xi} \right\rangle \, , \tag{16}$$

we obtain

$$\dot{\xi} = c(\xi) a(\xi, w) + b(\xi, w) \stackrel{\text{def}}{=} \theta(\xi, w) \tag{17}$$

with

$$a(\xi, w) = \left(1 - \frac{\left\langle v^\xi_\xi, p^\xi_\xi \right\rangle}{\left\langle p^\xi_\xi, p^\xi_\xi \right\rangle}\right) \Big/ \left(1 - \frac{\left\langle w, p^\xi_{\xi\xi} \right\rangle}{\left\langle p^\xi, p^\xi_\xi \right\rangle}\right) \, ,$$

$$b(\xi, w) = \frac{\left\langle \varepsilon^2 w_{xx} + f(p^\xi + w) - f(u^\xi) - f'(u^\xi) v^\xi, p^\xi_\xi \right\rangle}{\left\langle p^\xi_\xi, p^\xi_\xi \right\rangle \left(1 - \frac{\left\langle w, p^\xi_{\xi\xi} \right\rangle}{\left\langle p^\xi_\xi, p^\xi_\xi \right\rangle}\right)} \, .$$

By projecting equation (20) onto $F^\xi = Q^\xi L^2(0,1)$ we get, after using again (12) (16) and (17) and simple manipulations.

$$w_t = A^\xi w + \theta(\xi, w) K^\xi w + h(\xi, w) \ , \tag{18}$$

where

$$A^\xi w = Q^\xi(\varepsilon^2 w_{xx} + f'(p^\xi) w) \ , \tag{19}$$

$$K^\xi w = -\frac{\left\langle w, p^\xi_{\xi\xi} \right\rangle}{\left\langle p^\xi_\xi, p^\xi_\xi \right\rangle} \, p^\xi_\xi \ , \tag{20}$$

$$\begin{aligned} h(\xi, w) = {} & Q^\xi(f(u^\xi + v^\xi) - f(u^\xi) - f'(u^\xi) v^\xi) + \\ & Q^\xi(f(p^\xi + w) - f(p^\xi) - f'(p^\xi) w) + c(\xi) Q^\xi u^\xi_\xi \ . \end{aligned} \tag{21}$$

The system of differential equation (17), (18) is equivalent to equation (1) in the tubular neighborhood of $\Gamma\rho$ where the smooth change of variable $u \to (\xi, w)$ is well defined. Notice that at each time t the solution $(\xi(t), w(t))$ satisfies the condition

$$w(t) \epsilon F^{\xi(t)}.$$

This is a consequence of the presence of the expression $\theta(\xi, w) K^\xi w$ in the right hand side of (18).

In the following we shall assume that the linear fibration $\mathcal{F}$ has been smoothly extended in such a way that for $\xi < \frac{\rho}{2}, F^\xi = F^\rho$, and for $\xi > 1 - \frac{\rho}{2}, F^\xi = F^{1-\rho}$. We shall also assume that the function θ, h in (17) (18) extend to this fibration in such a way that for $\xi \notin \Gamma_{\rho/2} : \theta(\xi, w) = 0, h(\xi, w) = 0$.

The following lemma gives bounds on the functions θ, h and on their Lipschitz constants.

Lemma 3. *The right hand side $\theta(\xi, w)$ is a smooth function of $(\xi, w) \epsilon \boldsymbol{R} \times C^1[0,1]$. Moreover the conditions*

$$\|w\|_{C^1} < C \ \exp(-\frac{\nu}{\varepsilon}\ \rho)\ , \left\langle w, p_\xi^\xi \right\rangle = 0\ ,$$

imply

$$|\theta(\xi, w)| < C \ \exp\left(-\frac{2\nu}{\varepsilon}\ \rho\right)\ ,$$
$$|\theta_\xi(\xi, w)| < C \ \exp\left(-\frac{2\nu}{\varepsilon}\ \rho\right)\ ,$$
$$|\theta_w(\xi, w)\phi| \leq C \ \exp\left(-\frac{\nu}{\varepsilon}\ \rho\right) \|\phi\|_{C^1}\ ,$$

for some constant C independent of ξ, w, ε.

In the statement of Lemma 3 and in several other places in the following we use the symbol C or M or μ to denote a constant which may not be always the same.

Lemma 4. *The expression $h(\xi, w)$ on the right hand side of (18) defines a smooth function $h((\cdot,\cdot)) : \boldsymbol{R} \times W^{1,2}$ into L^2. Moreover the condition*

$$\|w\|_{W^{1,2}} < C \ \exp\left(-\frac{\nu}{\varepsilon}\rho\right)\ ,$$

implies

$$\|h(\xi, w)\|_{L^2} < C \ \exp\left(-\frac{2\nu}{\varepsilon}\rho\right)\ ,$$
$$\|h_\xi(\xi, w)\|_{L^2} < C \ \exp\left(-\frac{2\nu}{\varepsilon}\rho\right)\ ,$$
$$\|h_w(\xi, w)\phi\|_{L^2} < C \ \exp\left(-\frac{\nu}{\varepsilon}\rho\right) \|\phi\|_{W^{1,2}}\ ,$$

for some constant C independent of ξ, w, ε.

POINT 2: Define a smooth family of linear homeomorphisms $\Phi(\xi, \overline{\xi}) : F^{\overline{\xi}} \rightarrow F^{\xi}$ identifying the fiber F^ξ with the fiber $F^{\overline{\xi}}$.

These maps are of central importance in the derivation of the estimates needed to show that system (17) and (18) has an invariant manifold. The basic use of these maps is for comparing values of a section $\xi \to \sigma^\xi \epsilon F^\xi$ at two different values $\xi, \overline{\xi}$.

Let us consider the equation

$$\phi_\xi = K^\xi \phi = -\frac{\left\langle \phi, p^\xi_{\xi\xi} \right\rangle}{\left\langle p^\xi_\xi, p^\xi_\xi \right\rangle} p^\xi_\xi \ , \tag{22}$$

this can be regarded as an ordinary differential equation in L^2. Since K^ξ is a smooth function of ξ and K^ξ is bounded in L^2, equation (22) defines a group $\Phi(\xi, \overline{\xi})$ in L^2. The restriction of $\Phi(\xi, \overline{\xi})$ to $F^{\overline{\xi}}$ that we again label $\Phi(\xi, \overline{\xi})$ is a map onto F^ξ. This follows from

$$\left\langle \phi, p^\xi_\xi \right\rangle_\xi = \left\langle K^\xi \phi, p^\xi_\xi \right\rangle + \left\langle \phi, p^\xi_{\xi\xi} \right\rangle = 0 \ .$$

We also note that $\Phi(\xi, \overline{\xi}) : F^{\overline{\xi}} \to F^\xi$ is an isometry because for $\phi \epsilon \overline{F}^\xi$ we have

$$\langle \phi, \phi \rangle_\xi = 2 \left\langle K^\xi \phi, \phi \right\rangle = 0 \ ,$$

and therefore $\|\Phi(\xi, \overline{\xi})\phi\|_{L^2} = \|\phi\|_{L^2}, \phi \epsilon F^{\overline{\xi}}$. In the following we shall need also an estimate of the norm of $\Phi(\xi, \overline{\xi})$ as a map from L^2 onto L^2 and also an estimate of the differnce $\Phi(\xi, \overline{\xi})\phi - \phi$.

Lemma 5. *Assume that $\xi, \overline{\xi}$ are in a compact interval I. Then there exist numbers $C, \mu > 0$ independent of $\xi, \overline{\xi}, \varepsilon$ and such that*

$$\|\Phi(\xi, \overline{\xi})\phi - \phi\|_{W^{2,2}} \leq C\, \varepsilon^{-\mu} |\xi - \overline{\xi}| \ \|\phi\|_{L^2} \ ,$$

$$\|\Phi(\xi, \overline{\xi})\phi\|_{L^2} \leq (1 + C\, \varepsilon^{-\mu} |\xi - \overline{\xi}|) \ \|\phi\|_{L^2} \ ,$$

Moreover, if $\phi \epsilon C^1$ or $\phi \epsilon W^{2,2}$, the second estimate holds with the C^1 or the $W^{2,2}$ norm instead of the L^2 norm.

The proof of this lemma is obtained by analyzing the system of differential equations for the new unknowns $\alpha\epsilon\boldsymbol{R}$, $\sigma\epsilon F^{\xi}$ that can be derived by (22) by setting $\phi = \alpha p^{\xi}_{\xi} + \sigma$.

POINT 3: Study the linear fibration equation

$$\phi_t = A^{\xi}\phi + \dot{\xi}K^{\xi}\phi + \sigma^{\xi} \;, \tag{23}$$

where $\xi \to \sigma^{\xi}\epsilon F^{\xi}$ is a smooth section.

This study is based on Proposition 2 and aims to obtain exponential estimates analogous to the ones needed in standard invariant mainfold theory. The meaning of equation (23) is as follows: through equation (23) we associate to any given smooth function $\xi(t)$ the equation

$$\phi_t = A^{\xi(t)}\phi + \dot{\xi}(t)K^{\xi(t)}\phi + \sigma^{\xi(t)} \;. \tag{24}$$

this equation is the analogous of a linear inhomogeneous equation in a Banach space in the present setting of linear fibration. We can also consider equation (24) as an ordinary differential equation on L^2. From this observattion and standard theory [**He**] it follows that equation (24) with $\sigma = 0$ generates a linear semigroup $S(t,s,\xi(\cdot))$ on L^2 or on the fractional space $X^{\alpha} \subset L^2$ associated with the operator $A\phi = \varepsilon^2\phi_{xx}, D(A) = \{\phi|\phi\epsilon W^{2,2}, \phi_x(0) = \phi_x(1) = 0\}$. We shall assume $\alpha > \frac{3}{4}$ so that we have the imbedding $X^{\alpha} \subset C^1$.

Notice that, due to the presence of the term $\dot{\xi}K^{\xi}\phi$ in equation (23), $\phi(s)\epsilon F^{\xi(s)}$ implies that $S(t,s,\xi(\cdot))\phi(s)\epsilon F^{\xi(t)}$. Therefore equation (24) with $\sigma = 0$ defines a map $S(t,s,\xi(\cdot)) : F^{\xi(s)} \to F^{\xi(t)}$ for $t \geq s$.

Lemma 6. *Let $\xi(\cdot) : \boldsymbol{R} \to \boldsymbol{R}$ be a C^1 function such that $|\xi(t)| < C, |\dot{\xi}(t)| < C\epsilon^{\chi}$, with $C,\ \chi > 0$. Then for χ sufficiently large, there exist positive numbers M, β, independent of*

t, s, ε and also independent of $\xi(\cdot)$ for $\xi(\cdot)$ such that the following estimates hold for $\varepsilon \ll 1$

$$\|S(t,s,\ \xi(\cdot))\phi\|_\alpha \le M\exp(-\beta(t-s))\ \|\phi\|_\alpha$$

$$\|S(t,s,\ \xi(\cdot))\phi\|_\alpha \le M(t-s)^{-\alpha}\ \exp(-\beta(t-s))\ \|\phi\|_{L^2}$$

To prove this lemma one first transforms equation (24) with $\sigma = 0$ to an equation on the fixed Banach space $F^{\xi(s)}$ by means of the maps $\Phi(\xi, \overline{\xi})$ and then applies the results in [**He**] for linear nonautonomous equations. Proposition 2 is the basis for the exponential estimates. The constant β in Lemma 6 is related to the constant C in Proposition 2.

From the exponential estimates in Lemma 4 it follows that if $\xi \to \sigma^\xi \epsilon F^\xi$ is a bounded section and if $\xi(\cdot)$ is a C^1 function satisfying the hypothesis of the lemma. Then equation (24) has a unique solution which exists and is bounded in $(-\infty, t]$,

$$\phi(t, \xi(\cdot)) = \int_{-\infty}^{t} S(t,s,\xi(\cdot))\sigma^{\xi(s)}\, ds\ .$$

In the remaining part of the proof we shall need to compare the operators $S(t,s,\xi(\cdot))$, $S(t,s,\overline{\xi}(\cdot))$ corresponding to two different functions $\xi(\cdot), \overline{\xi}(\cdot)$. This comparison is done through the maps $\Phi(\xi, \overline{\xi})$ and is contained in the following lemma.

Lemma 7. *Let $\overline{\xi}(\cdot), \xi(\cdot) : \boldsymbol{R} \to \boldsymbol{R}$ be C^1 functions satsifying the assumptions in Lemma 6 with χ sufficiently large and let $\phi \epsilon F^{\overline{\xi}(s)}$. Then there exist positive numbers M, μ, β independent of $s, t, \varepsilon, \xi(\cdot), \overline{\xi}(\cdot)$ and such that, for $t \ge s$*

$$\|[\Phi(\overline{\xi}, \xi, t)S(t,s,\xi(\cdot))\Phi(\xi, \overline{\xi}, s) - S(t,s,\overline{\xi}(\cdot))]\phi\|_\alpha \le$$
$$M\varepsilon^{-\mu}e^{-\beta(t-s)}\|\phi\|_{L^2} \int_s^t (t-r)^{-\alpha}(r-s)^{-\alpha}|\xi(r) - \overline{\xi}(r)|dr\ ,$$

where $\Phi(\xi, \overline{\xi}, t) := \Phi(\xi(t), \overline{\xi}(t))$.

The proof of this lemma is based on Lemma 5 and 6 and is contained in [**ABF**].

POINT 4: Obtain an invariant manifold for system (17) and (18) as a fixed point of a suitable contraction map.

Now we try to adapt the classical Liapounov's approach to invariant manifolds to our situation. Our arguments are modeled after [**He**]. Let Σ be the set of sections $\xi \to \sigma^\xi \epsilon F^\xi$ which satisfy the conditions

$$(25) \qquad \begin{cases} \|\sigma^\xi\|_\alpha \leq \exp\left(-\frac{\nu}{\varepsilon}\rho\right) = D \\ \|\Phi(\overline{\xi},\xi)\sigma^\xi - \sigma^{\overline{\xi}}\|_\alpha \leq \Delta|\xi - \overline{\xi}|, \quad \xi;\overline{\xi}\epsilon \boldsymbol{R} \end{cases}$$

for some Δ independent of ε. With the norm $|||\sigma||| = \sup_\xi \|\sigma^\xi\|_\alpha$ the set of sections $\xi \to \sigma^\xi \ \epsilon F^\xi$ which are bounded in X^α is a Banach space. The set Σ defined by (25) is a compact subset of this Banach space. Condition (25) is, in our linear fibration setting, the analogous of the usual Lipshitz condition in the setting of Banach spaces. For $\sigma \ \epsilon \ \Sigma$ let $\xi(t) = \xi(t;\eta,\sigma)$, be the solution of the differential equation

$$(26) \qquad \begin{cases} \dot{\xi} = \theta(\xi,\sigma^\xi) \ , \\ \xi(0) = \eta \ . \end{cases}$$

From Lemma 3 and the fact that θ vanishes for $\xi \notin \Gamma\rho$ it follows that $\xi(\cdot)$ is a C^1 function which satisfies the hypothesis of Lemma 6. Therefore there is a unique solution bounded in $(-\infty, t]$ to the linear inhomogneous equation

$$(27) \qquad w_t = A^{\xi(t)}w + \theta(\xi(t),\sigma^{\xi(t)})K^{\xi(t)}w + h(\xi(t),\sigma^{\xi(t)}) \ .$$

We define $(G\sigma)^\eta$ to be the value at $t = 0$ of this solution

$$(28) \qquad (G\sigma)^\eta = \int_{-\infty}^0 S(0,s,\xi(\cdot))h(\xi(s),\sigma^{\xi(s)})ds \ .$$

We now show that for ε sufficiently small equation (27) defines a map $G : \Sigma \to \Sigma$ and that this map is a contraction. Before describing this part of the proof of theorem 1 we remark that in equation (27) $\xi(s)$ and $\xi(\cdot)$ stand for $\xi(s;\eta,\sigma)$ and $\xi(\cdot;\eta,\sigma)$ and therefore the dependence of $(G\sigma)^\eta$ on σ,η is rather involved. In the following, to keep the notation as simple as possible, we shall write $S(s,\xi(\cdot))$ instead of $S(0,s,\xi(0))$ and simply $\Phi(\xi,\overline{\xi})$ instead of $\Phi(\xi,\overline{\xi},t)=\Phi(\xi(t),\overline{\xi}(t))$ being clear from the contest at what time the functions $\xi,\overline{\xi}$ must be computed. We shall use the notation ρ^- to denote a positive number $<\rho$.

Lemma 7. *Let $\xi(t)=\xi(t;\eta,\sigma)$ the solution of (26) and let $\overline{\xi}(t)=\xi(t;\overline{\eta},\overline{\sigma})$ be the solution of (26) with $\sigma=\overline{\sigma},\eta=\overline{\eta}$. Then, provided $\sigma,\overline{\sigma}\epsilon\Sigma$ and $t\le 0$, the following estimate holds*

$$|\xi(t)-\overline{\xi}(t)|\le \exp(-dt)\,(|\eta-\overline{\eta}|+|||\sigma-\overline{\sigma}|||)-|||\sigma-\overline{\sigma}|||\,,$$

where $d=C\;\exp\left(-\frac{\nu}{\varepsilon}\rho^-\right)$ for some $C>0$ independent of ε.

Lemma 8. *Let $h(\xi,w)$ be the function defined by equation (21). Then there is a number C independent of ε such that for any $\xi,\overline{\xi}\;\epsilon\;\mathbf{R}$ and for any $\sigma,\overline{\sigma}\;\epsilon\Sigma$,*

$$\|\Phi(\overline{\xi},\xi)h(\xi,\sigma^{\xi})-h(\overline{\xi},\sigma^{\overline{\xi}})\|_{L^2}\le d(|\xi-\overline{\xi}|+|||\sigma-\overline{\sigma}|||)\,.$$

with $d=C\;\exp\left(-\frac{\nu}{\varepsilon}\rho^-\right)$.

The proofs of these two lemma is fairly standard and it is based on Lemma 4 and 5.

Proposition 3. *Provided $\varepsilon>0$ is sufficiently small, equation (27) defines a map $G:\Sigma\to\Sigma$ which is a contraction.*

PROOF: From equation (27), Lemma 4 and 6 it follows

$$\|(G\sigma)^\eta\|_\alpha\le d^2M\int_{-\infty}^{0}(-s)^{-\alpha}e^{\beta s}ds \tag{28}$$

which proves $\|(G\sigma)^\eta\|_\alpha\le D$ if $\varepsilon>0$ is sufficiently small. To show that under the same assumption $G\sigma$ satisfies also the other conditions required for a section in Σ and at the same time to prove that G is a contraction on Σ we estimate the expression

$$(29) \qquad \Phi(\overline{\eta},\eta)(G\sigma)^{\eta} - (G\overline{\sigma})^{\overline{\eta}} =$$
$$\int_{-\infty}^{0} \left[\Phi(\overline{\eta},\eta)S(s,\xi(\cdot))h(\xi,\sigma^{\xi}) - S(s,\overline{\xi}(\cdot))h(\overline{\xi},\overline{\sigma}^{\overline{\xi}})\right] ds =$$
$$\int_{-\infty}^{0} \left[\Phi(\overline{\eta},\eta)S(s,\xi(\cdot))\Phi(\xi,\overline{\xi}) - S(s,\overline{\xi}(\cdot))\right] \Phi(\overline{\xi},\xi)h(\xi,\sigma^{\xi})ds +$$
$$\int_{-\infty}^{0} S(s,\overline{\xi}(\cdot)) \left[\Phi(\overline{\xi},\xi)h(\xi,\sigma^{\xi}) - h(\overline{\xi},\overline{\sigma}^{\overline{\xi}})\right] ds ,$$

where we have written simply $h(\xi,\sigma^{\xi})$ instead of $h(\xi(s),\sigma^{\xi(s)})$. Let I_1, I_2, denote the two integrals in the last expression in equation (29). Then it follows from lemmas 6, 5, 4, and 7 that for $\varepsilon \ll 1$

$$\|I_1\|_{\alpha} \leq d^2 M\varepsilon^{-\mu}(|\eta-\overline{\eta}| + |||\sigma-\overline{\sigma}|||) \int_{-\infty}^{0} e^{(\beta-d)s} \left(\int_{s}^{0} (-r)^{-\alpha}(r-s)^{-\alpha}\right) drds$$
$$\|I_2\|_{\alpha} \leq d\ M\varepsilon^{-\mu}(|\eta-\overline{\eta}| + |||\sigma-\overline{\sigma}|||) \int_{-\infty}^{0} e^{(\beta-d)s}(-s)^{-\alpha}ds \ .$$

therefore we have

$$\|\Phi(\overline{\eta},\eta)(G\sigma)^{\eta} - (G\overline{\sigma})^{\overline{\eta}}\|_{\alpha} \leq dM\varepsilon^{-\mu}(|\eta-\overline{\eta}| + |||\sigma-\overline{\sigma}|||) \ .$$

Since $d = C\ \exp\left(-\frac{\mu}{\varepsilon}\rho^{-}\right)$ this inequality with $\sigma = \overline{\sigma}$ shows that, for $\varepsilon \ll 1, G\sigma$ also satisfies the second condition for being an element of Σ. The same inequality for $\eta = \overline{\eta}$ shows G is a contraction on Σ. Therefore there is a fixed point $\hat{\sigma}\ \epsilon\Sigma$. That this is an invariant manifold for (17), (18) and therefore for (1) is standard.

To conclude the proof we note that from the discussion above it follows $\|\hat{\sigma}^{\xi}\|_{\alpha} \leq C\ \exp\left(-\frac{2\nu}{\varepsilon}\rho^{-}\right), \xi\epsilon\Gamma\rho$. Since this is true for any $\rho < \frac{1}{2}I$ we have

$$\|\hat{\sigma}^{\xi}\|_{\alpha} = C\ \exp\left(-\frac{2\nu}{\varepsilon}\delta^{\xi}\right) \ .$$

The estimate for the difference $\hat{\gamma} - \gamma$ follows from $\hat{\gamma}(\xi) = \theta(\xi,\hat{\sigma}^{\xi})$, from (30), Proposition 1 and Lemma 2 . ∎

Acknowledgement

It is a pleasure to express my gratitude to N. Alikakos and P. Bates for many interesting and constructive conversations during the preparation of this work.

References

[ABF] N. Alikakos, P. Bates, G. Fusco, *Slow Motion for the Cahn-Hilliard Equation in one space dimension*, Preprint.

[CF] G. Caginalp, P.C. Fife, *Elliptic problems involving phase boundaries satisfying a curvatiure condition*, I.M.A. Jour. Appl. Math., 38 (1987), 185-217.

[CP] J. Carr, R.L. Pego, *Metastable patterns in solutions of* $u_t = \varepsilon^2 u_{xx} - f(u)$,preprint.

[F] G. Fusco *On the explicit construction of a system of ODE which has the same dynamics as a scalar parabolic PDE*, Journal Differential Equations, 69 (1987), 85-110.

[FH] G. Fusco, J.K. Hale, *Slow motion manifolds, dormant instability and singular perturbations*, Dynamic and Diff. Equations, 16 (1989).

[FG] P.C. Fife, G.S. Gill, *The phase-field description of mu. zones*, Physica D., in press.

[G1] M.E. Gurtin, *On the two-phase Stefan problem with interfacd energy and entropy*, Arch. Rat. Mech. anal., 96 (1986), 199-241.

[G2] M.E. Gurtin, *On phase transitions with bulk, interfacial and boundary energy*, Arch. Rat. Mech. Anal., 36 (1986), 243-264.

[H1] J.K. Hale, *Asymptotic behavior of dissipative systems*, AMS monographs, No. 25 (1988).

[He] D. Henry, *Geometric theory of semilinear parabolic equaions*, Lect. Notes in Math, Vol. 840, Springer-Verlag.

[NF] Y. Nishiura and H. Fujii, *SLEP method and the stability of singularly perturbed solutions with multiple internal layers in reaction-diffusion systems*, NATO ASI Vol. F37, Dynamics of Infinite Dimensional Systems, Ed. S.N. Chow and J.K. Hale (1978).

ON THE ISOTHERMAL MOTION OF A PHASE INTERFACE.

Morton E. Gurtin
Department of Mathematics
Carnegie Mellon, Pittsburgh PA 15213

1. INTRODUCTION.

A recent series of papers (Gurtin [1-3], Angenent and Gurtin [4]) began an investigation whose goal is a thermomechanics of two-phase continua based on Gibbs's notion of a sharp phase-interface endowed with thermomechanical structure. In [1] a new balance law, balance of capillary forces, was introduced and then applied in conjunction with suitable statements of the first two laws of thermodynamics; the chief results are thermodynamic restrictions on constitutive equations, exact and approximate free-boundary conditions at the interface, and a hierarchy of free-boundary problems. The simplest versions of these problems (the Mullins-Sekerka problems) are essentially the classical Stefan problem with the free-boundary condition $u=0$ for the temperature replaced by the condition $u=hK$, where K is the mean curvature of the free-boundary and $h>0$ is a material constant. This dependence on curvature renders the problem difficult, and apart from numerical studies involving linearization-stability, there are almost no supporting theoretical results.

For perfect conductors the theory is far more tractable; there the temperature is constant, and the underlying free-boundary problem reduces to a single set of evolution equations for the interface. The paper [4] develops further this theory of perfect conductors for interfaces that evolve as curves in $\mathbb{R}^2$.

It is the purpose of this review to discuss the chief results of [1] and [4]. For convenience, we follow [4] and restrict attention to the isothermal evolution of interfacial curves in $\mathbb{R}^2$.

2. BASIC LAWS.

We consider a body which occupies all of $\mathbb{R}^2$ and consists of two phases separated, at each time t, by a smooth *interface* $\mathcal{s}(t)$

which evolves smoothly in time. We assume that $\mathcal{C}(t)$ is closed, write $\Omega(t)$ for the *bounded* region enclosed by $\mathcal{C}(t)$, refer to the phase occupying $\Omega(t)$ as the *reference phase*, write $\mathbf{N}$ for the outward unit normal to $\partial\Omega(t) = \mathcal{C}(t)$, and choose the unit tangent $\mathbf{T}$ such that $\{\mathbf{T},\mathbf{N}\}$ is a positively-oriented basis of $\mathbb{R}^2$. Generally, we will consider $\mathbf{T} = \mathbf{T}(\theta)$ and $\mathbf{N} = \mathbf{N}(\theta)$ as functions of the angle θ from a fixed coordinate axis to $\mathbf{N}$. We let s denote arc length on $\mathcal{C}(t)$ (with s increasing in the direction of $\mathbf{T}$), and write $K = \mathbf{N} \cdot \mathbf{T}_s$ for the *curvature* of $\mathcal{C}(t)$, V for the *normal velocity* of $\mathcal{C}(t)$ in the direction $\mathbf{N}$.

The micromechanics of the interface is described by two functions: $\mathbf{C}(s,t)$, the force within $\mathcal{C}(t)$; $\mathbf{b}(s,t)$, the force exerted on $\mathcal{C}(t)$ per unit length. $\mathbf{C}(s,t)$ is the *capillary force;* if we write

$$\mathbf{C} = \sigma\mathbf{T} + \xi\mathbf{N}, \tag{2.1}$$

then $\sigma(s,t)$ is the *surface tension,* $\xi(s,t)$ is the *surface shear.*

We refer to the normal component b of $\mathbf{b}$ as the *normal interaction;* b represents the normal force exerted on the interface by the bulk material. Motion *tangential* to the interface depends on the choice of parameterization and is hence irrelevant to the physics; the intrinsic evolution of the interface is *normal* to itself. As is consistent with a "constraint" of this type, we leave the tangential component of b as *indeterminate.*

Balance of capillary forces has the local form ([1], eqt. (3.3))

$$\mathbf{C}_s + \mathbf{b} = \mathbf{0}, \tag{2.2}$$

an equation with normal component

$$\xi_s + \sigma K + b = 0. \tag{2.3}$$

We associate with each interfacial motion an *interfacial energy* $f(s,t)$ per unit length. In addition, the individual phases possess bulk energies; in accord with our tacit assumption of isothermal conditions, we assume that the energy of each phase is constant, and we write F for the *energy of the reference phase* minus that of the other phase.

The second law for any subregion R of the body is the assertion that the rate at which the energy increases plus the energy outflow cannot be greater than the power supplied to R. This global statement of the second law leads to the *energy inequality* ([1], eqt. (3.16))

$$f^{\circ} - \xi\theta^{\circ} + (\sigma - f)KV + (b + F)V \leq 0, \tag{2.4}$$

where, for any function $g(s,t)$, g° represents the normal time derivative of g following the interface.

3. CONSTITUTIVE EQUATIONS. THERMODYNAMIC RESTRICTIONS.

As *constitutive equations* we allow the energy, capillary force, and normal interaction to depend on the orientation of the interface through a dependence on θ, and on the kinetics of the interface through a dependence on V:

$$f = f(\theta,V), \qquad \mathbf{C} = \mathbf{C}(\theta,V), \qquad b = b(\theta,V). \tag{3.1}$$

The first two relations characterize the interface, the last models the interaction between the interface and the bulk material.

We require that all interfacial motions related through the constitutive equations (3.1) be consistent with the energy inequality (2.4). This leads to the following set of thermodynamic restrictions ([1], eqts. (4.5)-(4.7)):

(i) *the energy and capillary force are independent of the normal velocity;*

(ii) *the energy generates the capillary force through the relation*

$$\mathbf{C}(\theta) = f(\theta)\mathbf{T}(\theta) + f'(\theta)\mathbf{N}(\theta); \tag{3.2}$$

(iii) *the normal interaction is given by a relation of the form*

$$b(\theta,V) = -F - \beta(\theta,V)V, \qquad \beta(\theta,V) \geq 0. \tag{3.3}$$

Trivially, (3.2) implies that $\sigma(\theta) = f(\theta)$, $\xi(\theta) = f'(\theta)$.

We assume that $\beta(\theta,V)$ is independent of V and that $f(\theta),\beta(\theta) > 0$; the second of which ensures that the interactive force $-\beta(\theta)V$ oppose interfacial motion.

4. EVOLUTION EQUATIONS.

Balance of capillary forces (2.3) in conjunction with the reduced constitutive equations (3.2), (3.3) lead to an evolution equation which relates the normal velocity V to the curvature K:

$$\beta(\theta)V = [f(\theta) + f''(\theta)]K - F \tag{4.1}$$

([1], eqt. (8.5), [4] eqt. (5.7); see also Brakke [5], Gage [6,7], Gage and Hamilton [8], and Grayson [9], who discuss the equation $V=K$). The relation (4.1), when combined with purely kinematical conditions for an evolving curve lead to a system of evolution equations for the interface. For a convex section of the interface, this system reduces to a single partial differential equation for the velocity $V=V(\theta,t)$ ([4], eqt. (5.7)):

$$\Phi(\theta)V_t = [V + \Psi(\theta)]^2[V_{\theta\theta} + V], \tag{4.2}$$

$$\Phi(\theta) = [f(\theta) + f''(\theta)]/\beta(\theta), \qquad \Psi(\theta) = F/\beta(\theta).$$

For $\Phi(\theta)>0$ this equation is parabolic (aside from the trivial degeneracy $V=-\Psi(\theta)$ at inflection points) and yields a theory similar in structure to its isotropic counterpart based on $V=K-F$. There is, however, no compelling physical reason to exclude energies $f(\theta)$ for which $f(\theta)+f''(\theta)<0$ over ranges of the angle θ; in fact, material scientists often consider such ranges (Gjostein [10], Cahn and Hoffman [11]). For $f(\theta)+f''(\theta)<0$, (4.2) is backward-parabolic and corresponding evolution problems are generally not well posed. A necessary condition for the *statical* stability of the interface is that $f(\theta)+f''(\theta)\geq 0$; accordingly, we use the terms *strictly stable* or *unstable* according as $f(\theta)+f''(\theta)>0$ or $f(\theta)+f''(\theta)<0$.

5. STABLE ENERGIES ([4], §6,§7).

Consider now interfacial energies that are *strictly stable*. Steady solutions of (4.1) for which the interface is convex and infinite, in the shape of a bump, are deduced in [4]. The bump recedes in one solution and advances in the other; for the receding bump the kinetic coefficient can be arbitrary, but the advancing bump requires a nonconvex polar diagram for $\beta(\theta)$.

[4] analyzes the global behavior of a *smooth* interface as measured by its perimeter $L(t)$ and enclosed area $A(t)$. The main result is most easily stated in terms of a bounded solid in an infinite liquid bath:

$$\left.\begin{array}{l}\text{If the bath is not supercooled, then } A(t)\to 0; \text{ if the} \\ \text{bath is supercooled, then initially small interfaces have} \\ L(t)\to 0, \text{ initially large interfaces have } A(t)\to\infty.\end{array}\right\} \quad (5.1)$$

In addition, for the case in which $A(t)\to\infty$, the isoperimetric ratio $L(t)^2/4\pi A(t)$ remains bounded as $t\to\infty$. It is shown further that if (for a nonconvex interface) one defines a finger as a section of the interface between inflection points, then the total number of fingers as well as the total curvature of each finger cannot increase with time. These results presume the existence of a smooth, simple (non self-intersecting) interface. In this regard, it is clear that in certain circumstances the interface can pierce itself as it evolves.

6. UNSTABLE ENERGIES ([4], §8,§9).

[4] next considers energies $f(\theta)$ which are unstable for certain values of θ. Here it is convenient to introduce a global definition of stability based on ideas of Wulff [12], Herring [13], and Frank [14]. Global stability is defined in terms of the convexity of the Frank diagram, which is the polar diagram of the reciprocal function $f(\theta)^{-1}$; the convex sections of this diagram are referred to as the *globally stable sections,* the remaining sections as the *globally unstable sections.* These definitions are consistent: $f(\theta)$ is stable on globally stable sections; $f(\theta)$ is unstable somewhere within each globally unstable section.

One way of treating unstable energies is to allow the interface to be *nonsmooth* with corners that correspond to jumps in θ across globally unstable sections. Balance of capillary forces for corresponding "weak solutions" of the evolution equations leads to the requirement that $C(\theta)$ be continuous across each such corner; this requirement is automatically met when the corners are as specified above.

In contrast to standard results for a strictly stable energy, the presence of corners leads to the possibility of *facets* (flat sections); in fact, to the presence of *wrinklings,* where a wrinkling is a series of facets with normals that oscillate between two fixed values. Such wrinklings are dynamically stable: the lengths of the individual facets do not increase with time.

The use of corners leads to *free-boundary* problems for the evolution of the interface, as the positions of the corners are not generally known *a-priori*. The free-boundary conditions are similar in nature to those of the classical Stefan problem ([4], eqt. (9.17)). For the special case in which $\Omega(t)$ (and hence $\mathcal{C}(t)$) is strictly convex the angle θ can be used as the independent spatial variable. This greatly simplifies the problem as the angle-pairs that define the corners are known *a-priori*; they are the angle-pairs that bound the globally unstable sections of the Frank diagram. The underlying problem then consists in solving (4.2) away from the corners with the stipulation that

$$V\mathbf{N} - V_{\theta}\mathbf{T} \quad \text{be continuous across each corner.}$$

7. NONSMOOTH ENERGIES ([4], §10,§11).

Material scientists often consider interfacial energies that are continuous but have derivatives which suffer jump discontinuities (Herring [13,15], Cahn and Hoffman [11]). Such interfaces are studied in [4]; as before, corners are used to remove the globally unstable sections. In agreement with statical results, discontinuities in $f'(\theta)$ lead to facets in the evolving interface. The result (5.1) remains valid for nonsmooth, nonstable energies.

Following Taylor's [16] statical treatment of crystal shapes, [4] considers *crystalline energies,* for which the globally stable sections are isolated points (that is, for which the Frank diagram touches the boundary of its convex hull only at discrete points). An interesting property of crystalline energies is that their evolution is governed by a system of *ordinary differential equations* of a particularly simple form, involving only nearest-neighbor interactions. These equations are solved for a rectangular crystal; the corresponding solution shows that, in situations for which the crystal shrinks (cf. (5.1)), the corresponding *isoperimetric ratio generally tends to infinity,* in sharp contrast to an isotropic interface, which shrinks to a round point (Gage [6], Gage and Hamilton [8], Grayson [9]).

ACKNOWLEDGMENT. This work presented here was supported by the Army Research Office and the National Science Foundation.

REFERENCES.

[1] Gurtin, M. E., Multiphase thermomechanics with interfacial structure. 1. Heat conduction and the capillary balance law. Arch. Rational Mech. Anal. **104**, 195-221 (1988).

[2] Gurtin, M. E., On the two-phase Stefan problem with interfacial energy and entropy, Arch. Rational Mech. Anal. **96**, 199-241 (1986).

[3] Gurtin, M. E., Toward a nonequilibrium thermodynamics of two phase materials, Arch. Rational Mech. Anal. **100**, 1275-312 (1988).

[4] Angenent, S. and M. E. Gurtin, Multiphase thermomechanics with interfacial structure. 2. Evolution of an isothermal interface. Arch. Rational Mech. Anal., forthcoming.

[5] Brakke, K. A., *The Motion of a Surface by its Mean Curvature,*

[6] Gage, M., Curve shortening makes convex curves curcular, Invent. Math. **76**, 357-364 (1984)

[7] Gage, M., On an area preserving evolution equation for plane curves, Contemp. Math. **51**, 51-62 (1986)

[8] Gage, M. and R. S. Hamilton, The heat equation shrinking convex plane curves, J. Diff. Geom. **23**, 69-96 (1986).

[9] Grayson, M. A., The heat equation shrinks embedded plane curves to points, J. Diff. Geom. **26**, 285-314 (1987).

[10] Gjostein, N. A., Adsorption and surface energy 2: thermal faceting from minimization of surface energy, Act. Metall. **11**, 969-978 (1963).

[11] Cahn, J. W. and D. W. Hoffman, A vector thermodynamics for anisotropic surfaces. 2. curved and faceted surfaces, Act. Metall. **22**, 1205-1214 (1974).

[12] Wulff, G., Zur Frage der Geschwindigkeit des Wachsthums und der Auflosung der Krystallflachen, Zeit. Krystall. Min. **34**, 449-530 (1901).

[13] Herring, C., Some theorems on the free energies of crystal surfaces, Phys. Rev. **82**, 87-93 (1951).

[14] Frank, F. C., The geometrical thermodynamics of surfaces, *Metal Surfaces: Structure, Energetics, and Kinetics,* Am. Soc. Metals, Metals Park, Ohio (1963).

[15] Herring, C., Surface tension as a motivation for sintering, *The Physics of Powder Metallurgy* (ed. W. E. Kingston) McGraw-Hill, New York (1951).

[16] Taylor, J. E., Crystalline variational problems, Bull. Am. Math. Soc. **84**, 568-588 (1978).

For Additional references see the Appendix

THE RELAXED INVARIANCE PRINCIPLE AND WEAKLY DISSIPATIVE INFINITE DIMENSIONAL DYNAMICAL SYSTEMS

M. Slemrod[#]

Center for the Mathematical Sciences
University of Wisconsin
Madison, WI 53706

Abstract: This report outlines the role of the weak topology and the representation of weak limits by Young measures in proving the trend of dissipative infinite dimensional dynamical systems to equilibria. Two examples are provided: (i) a weakly damped wave equation and (ii) an infinite system of ordinary differential equations which model a liquid–vapor phase transition.

0. Introduction

Recently the subject of the dynamics of infinite dimensional dynamical systems has come in for renewed attention. For example two new books have appeared devoted to this subject. i.e. the monographs of Jack Hale [1] and Roger Temam [2]. In both of these books and most of the papers devoted to the subject, the dynamics are usually considered in the strong norm topology of the Banach space where the relevant flow exists. While this is natural for many applictions it excludes some interesting cases where the weak topology may be more useful.

[#]This research was supported in part by the Air Force Office of Scientific Research, Air Force Systems Command, USAF, under Contract/Grant No. AFOSR–87–0315. The United States Government is authorized to reproduce and distribute reprints for government purposes not withstanding any copyright herein.

Dedicated to Jack Hale on the occasion of his 60th birthday.

In two new papers [3,4] I have attempted to show how the weak topology combined with the representation expected values of weak limits in terms of Young measures yields a new view on the well known LaSalle Invariance Principle. (Zvi Artstein suggested the term "relaxed invariance principle" for the argument and that is the name I have been using.) In this report I will sketch how the relaxed invariance principle can be applied to two rather different problems. The first is the problem of a weakly damped wave equation while the second arises from considering an infinite system of ordinary differential equations which model the cluster dynamics of a liquid–vapor phase transition. Of course the interested reader should consult [3] and [4] for details.

1. A damped wave quation

Consider the problem of the asymptotic behavior as $t \to \infty$ of solutions of the damped wave equation

$$\frac{\partial^2 u}{\partial t^2} - \Delta u = -a(x)q(u_t) \quad \text{on } \mathbb{R}^+ \times \Omega , \tag{1.1}$$

$$u = 0 \quad \text{on } \mathbb{R}^+ \times \partial\Omega , \tag{1.2}$$

where Ω is an open, connected domain in $\mathbb{R}^N$, $N \geq 1$, with smooth boundary and $a \in L^\infty(\Omega)$, $a \geq 0$ a.e. in Ω. To guarantee that there is actually some region of damping we set

$$E = \{x \in \Omega; \, a(x) > 0\}$$

and assume

$$\text{meas}(E) > 0 .$$

Since we desire existence and uniqueness of solution of (0.1), (0.2) with initial data

$$\left.\begin{array}{l} u(x,0) = u_0(x) , \\ u_t(x,0) = u_1(x) , \end{array}\right\} \quad x \in \Omega \tag{1.3}$$

with $U_0 = [u_0,u_1] \in H = H_0^1(\Omega) \times L^2(\Omega)$ we shall also assume q is globally Lipschitz continuous, $q(0) = 0$. In this case (1.1), (1.2), (1.3) possesses a globally defined unique weak solution $U = [u,u_t] \in H$.

Multiplication of (1.1) by $\frac{\partial u}{\partial t}$ and application of Green's theorem yields the identity

$$\|U(t;U_0)\|^2 - \|U_0\|^2 = -2\int_0^t (a(\cdot)q(u_t),u_t)ds \tag{1.4}$$

where $(,)$ denotes the usual $L^2(\Omega)$ inner product and

$$(U,V)_H = (\nabla u,\nabla v) + (u_t,v_t),\ \|U\|^2 = (U,U)_H\,.$$

for $U = [u,u_t]$, $V = [v,v_t]$,

It is thus a trivial observation that the "energy" $\|U(t)\|^2$ is dissipated if $q(\xi)\xi \le 0$ for all $\xi \in \mathbb{R}$ and q is not identically zero. We can then ask the question: what additional restrictions should be placed on q to guarantee asymptotic decay to the $u \equiv 0$ equilibrium? Previous work on this problem by Dafermos [5] and Haraux [6] has shown that monotonicity in q suffices but monotonicity clearly is much more restrictive than one would expected is needed.

In [3] I outlined how a "relaxed invariance principle" can be used to obtain information on the asymptotic behavior of (1.1)–(1.3). Here I sketch the basic and rather straightforward ideas. First I state the main result.

Theorem 1. Assume $a \in L^\infty(\Omega)$ with meas $(E) > 0$, g is a globally defined Lipschitz continuous function: $\mathbb{R} \to \mathbb{R}$. Assume $\{\xi \in \mathbb{R};\ q(\xi) = 0\}$ is contained in either $(-\infty,0]$ or $[0,\infty)$. Then any weak solution $U(t;U_0) = \lfloor u,u_t \rceil$ in H for data $U_0 \in H$ converges weakly to 0 in H as $t \to \infty$.

The main idea of the proof is to ascertain the dynamics of (1.1)–(1.3) on the weak ω–limit set. Here the weak ω–limit set is defined to be $\omega(U_0)=\{[y_0,y_1] \in H;$ there exists a sequence $\{t_n\}$, $t_n \to \infty$ so that $U(t_n;U_0) \rightharpoonup [y_0,y_1]$ as $n \to \infty\}$ where $\rightharpoonup$ denotes weak convergence in H.

From the energy equality (1.9) we know the weak ω–limits set is non–empty.

So now fix $[y_0,y_1] \in \omega(U_0)$ and let $\{t_n\}$ be such that $U(t_n;U_0) \rightharpoonup [y_0,y_1]$ as $n \to \infty$. Consider the translation sequence

$$U_n(t) = U(t+t_n;U_0)\,.$$

Since the U_n are also solutions of (1.1)–(1.2) satisfying the initial conditions $U_n(0) = U(t_n;U_0)$ the energy inequality

$$\|U_n(t)\|^2 - \|U(t_n;U_0)\|^2 = -2\int_0^t (a(\ )q(u_{n_t}),u_{n_t})ds \tag{1.5}$$

is satisfied. Hence

$$\|U_n(t)\| \le \|U(t_n;U_0)\| \le \|U_0\|$$

and trivially on any interval $[0,T]$ we have $\{U_n\}$ in a bounded subset of $L^2((0,T);H)$ which implies the existence of subsequences

$$\begin{aligned} \nabla u_n &\rightharpoonup \nabla v \text{ weakly in } L^2(Q)\,, \\ u_{n_t} &\rightharpoonup v_t \text{ weakly in } L^2(Q)\,, \end{aligned} \tag{1.6}$$

for some $[v,v_t] \in L^2((0,T);H)$, $Q = (O,T) \times \Omega$.

Using (1.6) we can now attempt to derive the dynamics on $\omega(U_0)$. First since

$$\frac{\partial^2 u_n}{\partial t^2} - \Delta u_n = -a(x)q(u_{n_t}) \text{ on } \mathbb{R}^+ \times \Omega\,, \tag{1.7}$$

$$u_n = 0 \text{ on } \mathbb{R}^+ \times \partial\Omega\,, \tag{1.8}$$

we can pass to the weak limit in (1.7), (1.8) to find

$$\frac{\partial^2 v}{\partial t^2} - \Delta v = \text{weak}\lim_{n\to\infty}(-a(x)q(u_{n_t})) \tag{1.9}$$

$$v = 0 \quad \text{on } \mathbb{R}^+ \times \partial\Omega . \tag{1.10}$$

Next since $\|U(t;U_0)\|$ is a nonincreasing function of t bounded from below by zero the limit of $\|U(t;U_0)\|$ must exist. Hence $\lim_{n\to\infty} \|U_n(t)\| = \lim_{n\to\infty} \|U(t_n;U_0)\|$ and from (1.5) we see

$$\lim_{n\to\infty} \int_0^T (a(\cdot)q(u_{n_t}),u_{n_t})\, ds = 0 . \tag{1.11}$$

Unfortunately we cannot pass our limits through q since we do not wish to assume any special weak sequential continuity of the map $q:L^2(\Omega) \to L^2(\Omega)$. Nevertheless there is a ploy which allows us to get around this difficulty. In fact it is precisely this step that may also be useful in other weekly damped problems.

First recall a result of M. Schonbek [7] on the representation of weak limits in terms of Young measures (see also Ball [8], Tartar [9], Young [10]).

<u>Proposition</u> Let O be an open set in $\mathbb{R}^m$. Let $w_n : O \mapsto \mathbb{R}^q$ be a sequence of functions uniformly bounded in $(L^p(O))^{1/q}$ for some $p > 1$. Then there exists a subsequence $\{w_{n_k}\}$ and a family of probability measures $\{\nu_y\}_{y\in O}$ on $\mathbb{R}^q$ so that if $f \in C(\mathbb{R}^q;\mathbb{R})$ and satisfies $f(w) = o(|w|^p)$ as $|w| \to \infty$ then

$$f(w_{n_k}) \rightharpoonup \langle \nu_y, f(\lambda)\rangle = \int_{\mathbb{R}^q} f(\lambda)d\nu_y(\lambda) \tag{1.12}$$

in the sense of distributions. Furthermore if O is bounded the above convergence is in $L^1(O)$ weak. (Here $f(w) = o(|w|^p)$ means $|f(w)|/|w|^p \to 0$ as $|w| \to \infty$.) Now we can apply the proposition to the sequence $u_{n_t} \in L^2(Q)$. Thus there exists a propability measure $\nu_{x,t}$ such that

$$q(u_{n_t}) \rightharpoonup \langle q(\lambda), \nu_{x,t}\rangle \tag{1.13}$$

weakly in $L^1(Q)$. Since $\lambda q(\lambda)$ is not necessarily $o(|\lambda|^2)$ we cannot apply the proposition direction to $\lambda q(\lambda)$. However we can use the following trick suggested by E. Zuazua. Define

$$g(\lambda) = \lambda q(\lambda), \quad |\lambda| \le 1,$$

$$= |q(\lambda)|, \quad |\lambda| > 1.$$

Then $\lambda q(\lambda) \ge g(\lambda) \ge 0$, $g(\lambda) = o(|\lambda|^2)$ and (1.11) implies

$$\lim_{n\to\infty} \int_0^T \int_\Omega a(x) g(u_{n_t}) ds = 0. \tag{1.14}$$

In addition we now have

$$g(u_{n_t}) \rightharpoonup \langle g(\lambda), v_{x,t} \rangle \tag{1.15}$$

weakly in $L^1(Q)$.

We now pass to the limits in (1.9) and (1.14) using the representations (1.13) and (1.15). Hence we see

$$\frac{\partial^2 v}{\partial t} - \Delta v = -a(x) \langle g(\lambda), v_{x,t} \rangle \text{ in } \Omega, \tag{1.16}$$

$$v = 0 \text{ on } \partial\Omega,$$

$$v(0,x) = y_0(x),$$

$$v_t(0,x) = y_1(x),$$

and

$$\langle g(\lambda), v_{x,t}(\lambda) \rangle = 0 \tag{1.17}$$

for almost all x,t, $x \in E$, $0 \le t \le T$. But (1.16) implies that

$$\operatorname{supp} \nu_{x,t} \subseteq \{\xi \in \mathbb{R};\ q(\xi) = 0\}$$

for almost all x,t, $x \in E$, $0 \le t \le T$. Hence from our hypothesis on q

$$\operatorname{supp} \nu_{x,t} \subseteq (-\infty,0] \quad \text{or} \quad \operatorname{supp} \nu_{x,t} \subseteq [0,\infty)$$

for almost all x,t, $x \in E$, $0 \le t \le T$. But this implies that $\frac{\partial v}{\partial t}$ is either non–positive or non–negative a.e in $E \times [0,T]$ while from (1.16) and (1.17) we see v satisfies the undamped wave equation

$$\frac{\partial^2 v}{\partial t^2} - \Delta v = 0 \text{ in } \mathbb{R}^+ \times \Omega$$

$$v = 0 \text{ on } \mathbb{R}^+ \times \partial\Omega$$

$$v(0,x) = y_0(x)\,,$$

$$v_t(0,x) = y_1(x)\,.$$

Finally an argument of C. M. Dafermos [5] shows that the only v satisfying the undamped wave equation with above sign constraint on $\frac{\partial v}{\partial t}$ is $v \equiv 0$. Hence $y_0 = y_1 = 0$ and $\omega_w(U_0) = [0,0]$.

2. Dynamics of the Becker–Döring cluster equations.

Let $c_r(t) \ge 0$ denote the number of r–particle clusters (or droplets) per unit volume at time t in a condensing vapor. In the vapor clusters can coagulate to form larger clusters or fragment to form smaller ones. An infinite set of ordinary differential equations describing the dynamics of these clusters was proposed by Lebowitz and Penrose [11] modifying an earlier model of Becker and Döring [12]:

$$\dot{c}_r(t) = J_{r-1}(c) - J_r(c), \quad r \geq 1 ,$$
$$\dot{c}_1(t) = -J_1(c) - \sum_{r=1}^{\infty} J_r(c) , \tag{2.1}$$

where $c = (c_r)$. Here a_r, b_{r+1} are the kinetic coefficients which are non–negative constants and

$$J_r(c) \overset{\text{def.}}{=} a_r c_1 c_r - b_{r+1} c_{r+1} .$$

A general study of existence, uniqueness, continuous dependence of initial data, and the long time trend to equilibrium of (1.1) was given in the papers of Ball, Carr, and Penrose [13], and Ball and Carr [14]. In [13] the authors showed if $a_r = O(r)$ and $c(0)$ is such that $\sum_{r=1}^{\infty} rc_r(0) < \infty$ then (2.1) possesses a solution for all $t > 0$ which conserves the density $\rho = \sum_{r=1}^{\infty} rc_r(t)$ for all $t \in [0,\infty)$. Furthermore they showed in the cases of pure fragmentation $(a_r = 0, b_{r+1} > 0$ for all $r \geq 1)$ and pure coagulation $(a_r > 0, b_{r+1} = 0$ for all $r \geq 1)$ solutions converge strongly in the Banach space $X = \{c; \sum_{r=1}^{\infty} rc_r < \infty\}$, $\|c\| = \sum_{r=1}^{\infty} rc_r$, to an equilibrium solution c^ρ of (1.1), $\|c^\rho\| = \rho_0$, $\rho_0 = \sum_{r=1}^{\infty} rc_r(0)$.

In the case of fragmentation and coagulation $(a_r > 0, b_{r+1} > 0$ for all $r \geq 1)$ they obtained a particularly important result (Theorem 5.5 of [13]). Namely if $Q_r \overset{\text{def.}}{=} \prod_{\ell=2}^{r} (\frac{a_{\ell+1}}{b_\ell})$ and we assume

$$a_r = O(r/\ell n\ r),\ b_r = O(r/\ell n\ r) \tag{2.2}$$

then any solution $c(t)$ of (1.1) approaches an equilibrium c^ρ of (1.1) componentwise, i.e.

$$c_r(t) \to c_r^\rho \quad \text{as} \quad t \to \infty$$

for some $0 \le \rho \le \min(\rho_0, \rho_s)$. (Here ρ_s is the density of saturated vapor defined as $\rho_s = \sum_{r=1}^{\infty} rQ_r z_s^{\ell}$ and z_s is the radius of convergence of the series.) The result is particularly striking when the initial data $c(0)$ is such that $\rho_0 > \rho_s$. For in this case the initial density ρ_0 is greater than the saturation density ρ_s and convergence of c to c^ρ cannot occur in the strong (norm) topology of X. For if it did density conservation would imply $\rho = \rho_0$ and hence $\rho > \rho_s$, a contradiction to the fact that there are no equilibrium densities greater than ρ_s.

A natural interpretation of this result is to regard the coagulation–fragmentation dynamics when $\rho_0 > \rho_s$ as a dynamic phase transition. Namely we might regard the initial data $c(0)$ to be in a metastable set which evolves as $t \to \infty$ to a new equilibrium consisting of vapor with density $\rho \le \rho_s$ and a liquid phase. The occurence of a liquid phase which is not accounted for in the original equations (1.1) would explain the apparent lack of density conservation as $t \to \infty$.

The proof of Theorem 5.5 of [13] is based on "invariance principle" Lyapunov function argument. The crucial steps in applying the "invariance principle" in its usual form are have (i) precompactness of orbits of the relevant dynamical system in a suitable metric space, and have both (ii) continuity with respect to initial conditions and (iii) the existence of a continuous nonincreasing Lyapunov function (with respect to the same suitable metric space as (i)). In their paper [13], the authors showed (i) and (iii) followed naturally by use of the metrized weak* topology on the ball of radius ρ_0 in X. However to show (ii) they were called upon to introduce the extra assumption (1.2). In [4] I have shown that (1.2) can be replaced by the weaker hypothesis

$$a_r = O(r),\ b_r = O(r) \qquad (2.3)$$

and still obtain the trend to equilibria described above. Specifically the following theorem was proven.

Theorem 2. Assume $a_r > 0, b_r > 0$ for all r, $a_r = O(r), b_r = O(r)$ and that $\limsup_{r\to\infty} Q_r^{1/r} < \infty$. Let c be any solution of (2.1) with $V(c(0)) < \infty$ where

$$V(c) = \sum_{r=1}^{\infty} c_r\left(\ell n\left[\frac{c_r}{Q_r}\right] - 1\right)$$

and

$$\rho_0 = \sum_{r=1}^{\infty} rc_r(0).$$

Then there is an equilibrium state c^ρ so that $c(t) \to c^\rho$ weak* in X as $t \to \infty$ for some ρ with $0 \le \rho \le \min(\rho_0, \rho_s)$. If in addition $\lim_{r\to\infty} Q_r^{1/r}$ exists then ρ is unique for each such $c(0)$ independently of which solution c of (2.1) we may consider.

Sketch of proof. The flavor of the proof is very much reminiscent of the arguments given for Theorem 1. I sketch here the part on asymptotic decay to an equilibrium.

First we note that for any solution of (2.1) conservation of mass shows

$$\|c(t)\| = \sum_{r=1}^{\infty} rc_r(t) = \sum_{r=1}^{\infty} rc_r(0) < \infty$$

and hence any fixed positive orbit in X possesses a non–empty weak* ω–limit set $\omega^*(c(0))$. Fix an element $y \in \omega^*(c(0))$ and let $\{t_n\}$ be the associated sequence.

Now as in Theorem 1 we consider a translate sequence $c^{(n)}(t) = c(t+t_n;c_0)$ in X for $t \ge 0$. We then recall the following result of L. Tartar [9] on the representation of weak* limits in terms of Young measures.

Proposition Suppose K is a bounded set in $\mathbb{R}^m$. Let $w_n: \Omega \to \mathbb{R}^q$ be such that $w_n \in K$ a.e. Then there exists a subsequence $\{w_{n_k}\}$ and a family of probability measures $\{\nu_x\}_{x\in\Omega}$ on $\mathbb{R}^m$ with supp $\nu_x \subset \overline{K}$, such that if f is any continuous function as K, then

$$f(w_{n_k}) \to \langle \nu_x, f(\lambda)\rangle \text{ in } L^\infty(\Omega) \text{ weak}^*,$$

where

$$\langle \nu_x, f(\lambda)\rangle = \int_K f(\lambda) d\nu_x(\lambda) .$$

Since

$$0 \le \sum_{r=1}^{\infty} b_{r+1} c_{r+1}^{(n)} \le \text{const.} \sum_{r=1}^{\infty} (r+1) c_{r+1}^{(n)} \le \text{const.} \sum_{r=1}^{\infty} (r+1) c_{r+1}^{(n)} \le \text{const } \rho_0$$

and

$$0 \le \sum_{r=1}^{\infty} a_r c_1^{(n)} c_r^{(n)} \le \text{const.} \sum_{r=1}^{\infty} c_1^{(n)} c_r^{(n)} \le \text{const.}\ \rho_0^2 ,$$

the sequences

$$\{ \sum_{r=1}^{\infty} a_r c_1^{(n)} c_r^{(n)}, \sum_{r=1}^{\infty} b_{r+1} c_{r+1}^{(n)} \}$$

lie in a bounded set of K of $\mathbb{R}^+ \times \mathbb{R}^+$ for $0 < t < \infty$. Furthermore a little work shows these sequences are each uniformly bounded away from zero. So we can apply the Proposition on $L^\infty(0,\infty)$ to represent the weak* limits of there two sequences by a Young measure $\nu_t(\lambda_1,\lambda_2)$. Choosing additional sequences if necessary we then see there is a subsequence $\{c^{(n_k)}\}$ so that

$$c^{n_k} \to \bar{c} \text{ weak } * \ L^\infty(0,\infty)$$

and

$$\bar{c}_r(t) = y_r + \int_0^t [J_{r-1}(\bar{c}(s)) - J_r(\bar{c}(s))] ds, \qquad r \ge 2$$

$$\bar{c}_1(t) = y_1 - \int_0^t [J_1(\bar{c}(s)) + \langle v_s, \lambda_1 - \lambda_2 \rangle] ds \tag{2.4}$$

where we recall y is the fixed element of $\omega^*(c(0))$.

Now from differentiating the Lyapunov function $V(c(t))$ along trajectories we have

$$V(c(t)) + \int_0^t D(c(s)ds \subseteq V(c(0)) \quad \text{for all} \quad t \geq 0$$

where $D(c)$ is a non–negative quantity defined by

$$D(c) = \sum_{r=1}^{\infty} (a_r c_1 c_r - b_{r+1} c_{r+1})(\ell n(a_r c_1 c_r) - \ell n(b_{r+1} c_{r+1})) .$$

But by Jensen's inequality $D(c)$ is greater than or equal to the non–negatiave quantity

$$[\sum_{r=1}^{\infty} a_r c_1 c_r - \sum_{r=1}^{\infty} b_{r+1} c_{r+1}][\ell n[\sum_{r=1}^{\infty} a_r c_1 r_n] - \ell n[\sum_{r=1}^{\infty} b_{r+1} c_{r+1}]] .$$

(2.5)

We now argue as in Theorem 1, i.e.

$$V(c^{(n_k)}(t)) + \int_0^t D(c^{(n_k)}(s))ds \leq V(c^{(n_k)}(0))$$

and since $V(c(t))$ is a nonincreasing function bounded from below $\lim_{t \to \infty} V(c(t))$ exists. Therefore on any interval $(0,T)$

$$\lim_{n_k \to \infty} \int_0^T D(c^{(n_k)}(s))ds = 0$$

and hence

$$\int_0^T \langle \nu_s, (\lambda_1-\lambda_2)(\ln \lambda_1 - \ln \lambda_2)\rangle ds = 0$$

where we have used (2.5). Thus

$$\text{supp } \nu_t \subseteq \{\lambda \in \mathbb{R}^+ \times \mathbb{R}^+ : \lambda_1 = \lambda_2\} \cap [\delta_T, a] \times [\delta_T, b] \quad \text{a.e in} \quad (0,T)$$

for some positive constants a,b,δ_T. Hence $J_r(\bar{c}(t)) = 0$ for all $t \in (0,T]$ and from (2.4) we see $\bar{c}(t) = y$ for all $t \in (0,T]$. So the weak* ω–limit set consists only of equilibria.

3. Conclusion

Two examples demonstrating the role of the weak topology in the analysis of infinite dimensional dynamical systems have been given but I am sure there are many more. Such problems will naturally occur when the dissipative mechanisms are very mild and subtle and when a priori estimates are sparse. It is in these cases that the "relaxed invariance principle" will play a crucial role in the analysis of long time asymptotic behavior.

References

1. J. Hale, "Asymptotic Behavior of Dissipative Systems", American Mathematical Society, Providence, RI, 1988.

2. R. Temam, "Infinite–Dimensional Dynamical Systems in Mechanics and Physics", Applied Mathematical Sciences 68, Springer Verlag, New York–Berlin–Heidelberg, 1988.

3. M. Slemrod, Weak asymptotic decay via a "relaxed invariance principle" for a wave equation with nonlinear, nonmonotone damping, to appear Proc. Royal Society of Edinburgh.

4. M. Slemrod, Trend to equilibrium in the Becker–Döring cluster equations, to appear Nonlinearity.

5. C. M. Dafermos, Asymptotic behavior of some evolutionary systems, in "Nonlinear Evolution Equations" (M. G. Crandall, ed.) pp. 141–154, Academic Press, New York, 1978.

6. A. Haraux, Stabilization of trajectories for some weakly damped hyperbolic equations, J. Differential Equations 59 (1985), 145–154.

7. M. Schonbek, Convergence of solutions to nonlinear dispersive equations, Comm. in Partial Differential Equations 7 (1982), 959–1000.

8. J. M. Ball, A version of the fundamental theorem for Young measures, to appear in Proc. of CNRS–NSF Conference on Continuum Models of Phase Transitions, Springer Lecture Notes in Mathematics (ed. D. Serre, M. Rascle, M. Slemrod), 1990.

9. L. Tartar, Compensated compactness and applications to partial differential equations, in "Research Notes in Mathematics 39; Nonlinear Analysis and Mechanics: Heriot–Watt Symposium, Vol. IV "(R. J. Knops, ed.) pp. 136–211, Pitman Press, London, 1975.

10. L. C. Young, "Lectures on the Calculus of Variations and Optimal Control Theory", W. B. Saunders Co., Philadelphia, London, 1969.

11. O. Penrose and J. L. Lebowitz, Towards a rigorous molecular theory of metastability, in "Studies in Statistical Mechanics VII, Fluctuation Phenomena" (E. Montroll and J. L. Lebowitz, ed.), North Holland, 1976.

12. R. Becker and W. Döring, Kinetische Behandlung der Keimbildung in übersattigten Dämpfer, Ann. Phys. (Leipzig) 24 (1935), 719–752.

13. J. M. Ball, J. Carr, O. Penrose, The Becker–Döring cluster equations: basic properties and asymptotic behavior of solutions, Comm. Math. Physics 104 (1986), 657–692.

14. J. M. Ball and J. Carr, Asymptotic behavior of solutions to the Becker–Döring equations for arbitrary initial data, Proc. Royal Society of Edinburgh 108A (1988), 109–116.

3. Viscoelasticity

MATHEMATICAL PROBLEMS ASSOCIATED WITH THE ELASTICITY OF LIQUIDS

D. D. Joseph

Department of Aerospace Engineering and Mechanics
University of Minnesota, Minneapolis, MN 55455

This lecture is in three parts:

1. Physical phenomena associated with hyperbolicity and change of type;

2. Conceptual ideas associated with effective viscosities and rigidities and the origins of viscosity in elasticity;

3. Mathematical problems associated with hyperbolicity and change of type.

The ideas which I will express in this lecture are a very condensed form of ideas which have been put forward in various papers and most completely in my forthcoming book "Fluid Dynamics of Viscoelastic Liquids" which is to be published in 1989 by Springer-Verlag. The mathematical theory of hyperbolicity and change of type is associated with models with an instantaneous elastic response. Basically, this means that there is no Newtonian like part of the constitutive equation. The theory for these models as it is presently known is in my book. I am persuaded that further development of this subject lies in the realm of physics rather than mathematics. The main issues are centered around the idea of the effective viscosity and rigidity and the measurements of slow speeds, topics which are discussed in this paper in a rather more discursive than mathematical manner.

1. Physical phenomena associated with hyperbolicity and change of type

It is well known that small amounts of polymer in a Newtonian liquid can have big effects on the dynamics of flow. Drag reductions of the order of 80% can be achieved by adding polymers in concentrations of fifty parts per million to water. This minute addition does not change the viscosity of the liquid but evidently has a strong effect on other properties of the liquid which have as yet been inadequately identified.

We are going to consider some effects of adding minute quantities of polyethylene oxide to water on the flow over wires. The first experiments were on uniform flow with velocity U across small wires, flow over a cylinder. James and Acosta [1970] measured the heat transferred from three wires of diameter D=0.001, 0.002 and 0.006 inches. They used three different molecular weights of polymers in water (WSR 301, 205 and coagulant) in concentrations ϕ ranging from 7 parts to 400 parts per million by weight, the range of extreme dilution, in the drag reduction range. They found a critical velocity U_c in all cases except the case of most extreme dilution ϕ=6.62 ppm, as is shown in figure 1. A brief summary of the results apparent in this figure follows.

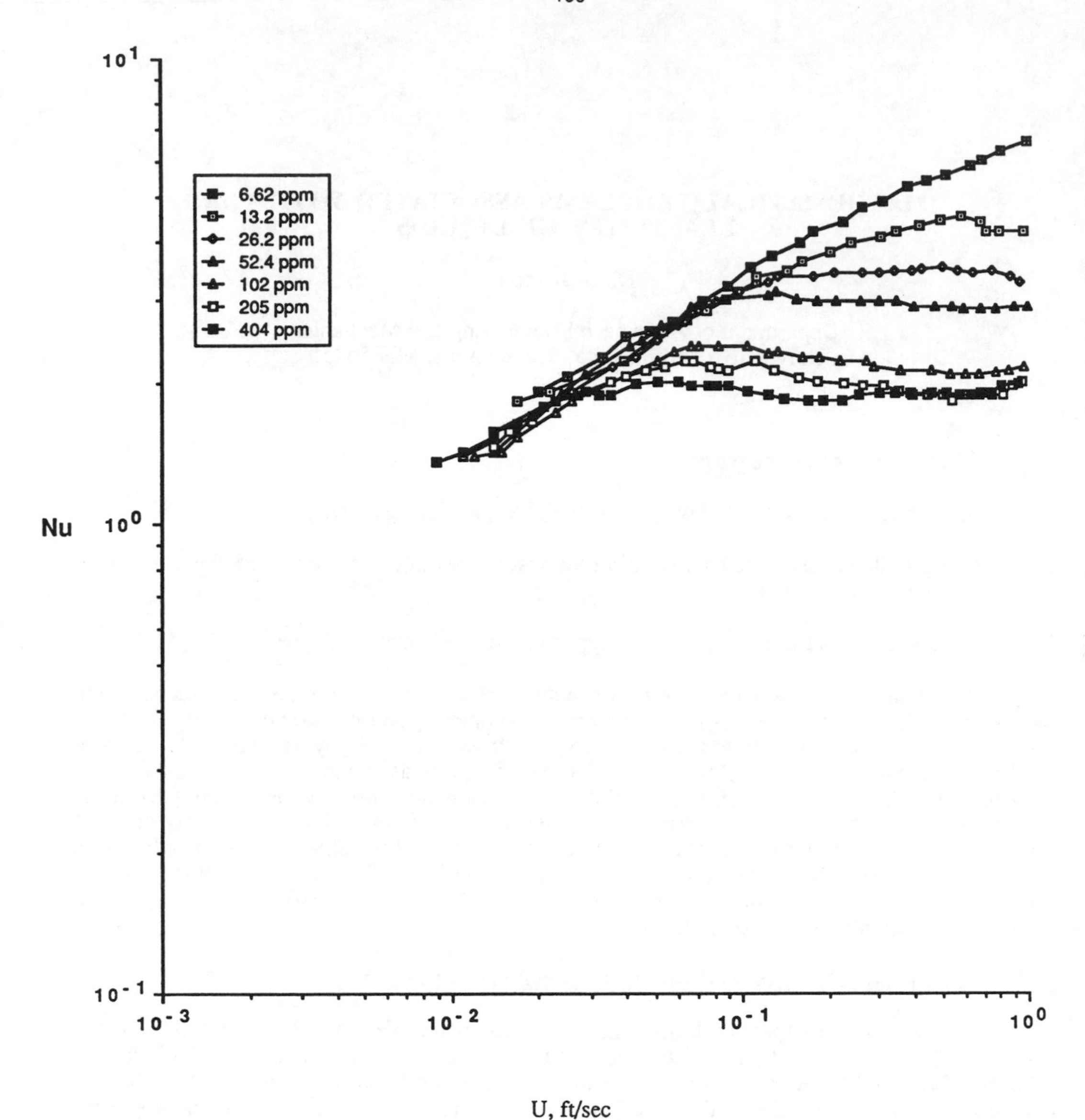

Figure 1. Heat transfer from heated wires in the cross-flow of WSR-301 (after James and Acosta, 1987). The critical U is independent of wire diameter.

1. There is a critical value U_c for all but the most dilute solutions: When $U<U_c$, the Nusselt number Nu(U) increases with U as in a Newtonian fluid. For $U>U_c$, the Nusselt number becomes independent of U as in figure 1.

2. U_c is independent of the diameter of the wire. This is remarkable. It suggests that U_c is a material parameter depending on the fluid alone.

3. U_c is a decreasing function of ϕ, the concentration. It is useful to note once again that in the range of ϕ between 6 ppm to 400 ppm, the viscosity is essentially constant and equal to the viscosity of water.

Ambari, Deslouis, and Tribollet [1984] obtained results for the mass transfer from 50 micron wires in a uniform flow of aqueous polyox (coagulant) solution in concentrations of 50, 100, and 200 parts per million. Their results are essentially identical to those obtained by James and Acosta [1970]; there is a critical U_c, a decreasing function of ϕ, signalling a qualitative change for the dependence of the mass transport of U, from a Newtonian dependence when $U<U_c$, to a U independent value for $U>U_c$. Their values of U_c for the break in the mass transport curve are just about the same as the value of U_c found by James and Acosta for heat transfer.

Ultman and Denn [1970] suggested that $U_c=c=\sqrt{\eta/\lambda\rho}$ where η is the viscosity, λ the relaxation time, and ρ is the density of a fluid whose extra stress $\tau=\mathbf{T}+p\mathbf{1}$ satisfies Maxwell's equation

$$\lambda U\partial\tau/\partial x + \tau = \mu[\nabla\mathbf{u} + \nabla\mathbf{u}^T] \tag{1}$$

where $\mathbf{u}$ is the velocity. They used the molecular theory of Bueche to find the value of the relaxation time λ_B for the 52.4 ppm solution and they found that a $0.7\lambda_B$ would give $\sqrt{\eta/0.7\lambda_B}$ $=U_c \sim 2.9$ cm/sec., that is, their estimate of λ_B from Bueche's theory is almost good enough to give $c=U_c$. Their calculation of the time of relaxation cannot be relevant, however, because in the Bueche theory λ_B does not go to zero with the concentration ϕ. The zero ϕ value of λ_B can be interpreted as a relaxation time for a single polymer in a sea of solvent. The relaxation time of one polymer cannot be the relaxation time of the solution in the limit in which the polymer concentration tends to zero, because in this limit the solution is all solvent.

Joseph, Riccius and Arney [1986] measured c=2.48 cm/sec in a 50 ppm, WSR 301 aqueous solution. This measurement supports the idea that $U_c=c$. We are trying now to measure wave speeds in extremely dilute solutions in the drag reduction range. We find considerable scatter in our data in these low viscosity solutions and are at present uncertain about the true value of the effective wave speed, including the values which we reported earlier.

The hypothesis that $U_c=c$ is consistent with the following argument about the dependence of the wave speed on concentration. In the regime of extreme dilution, the viscosity does not change with concentration. However, there appears to be a marked effect on the average time of relaxation which increases with concentration. It follows then that the wave speed $c=\sqrt{\eta/\rho\lambda}$ must decrease with concentration ϕ.

Konuita, Adler and Piau [1980] studied the flow around a 0.206 mm wire in an aqueous polyox solution (500 ppm, WSR-301) using laser-Doppler techniques. They found a kind of shock wave in front of the cylinder, like a bow shock. They say that the velocity of the fluid is zero in a region fluid in front of the stagnation point. Basically they say that there is no flow, or very slow flow near the cylinder. The formation of the shock occurs at a certain finite speed, perhaps U_c. This type of shock is consistent with the other observations in the sense that with a stagnant region around the cylinder, the transport of heat and mass could take place only by diffusion, without convection. This explains why there is no dependence of the heat and mass transfer on the velocity when it exceeds a critical value.

I estimated the critical speed, using the data of Konuita, Adler and Piau, and I estimated the wave speed c by extrapolating from our measurements in the polyox solutions at different concentrations. These estimates are reported in my new book "Fluid Dynamics of Viscoelastic Liquids." They are consistent with the notion of a supercritical shock transition at $U_c=c$.

Another striking phenomenon which appears to be associated with a supercritical transition is delayed die swell. It is well known that polymeric liquid will swell when extruded from small diameter pipes. The swelling can be very large, four, even five times the diameter of the jet. This swelling is still not well understood even when there is no delay. Joseph, Matta and Chen [1987] have carried out experiments on 19 different polymer solutions. They found that there is a critical value of the extrusion velocity U_c such that when $U<U_c$, the swell occurs at the exit, but when $U>U_c$ the swell is delayed, as in figure 2. If U is taken as the centerline velocity in the pipe, then the transition is always supercritical with $U_c>c$. The length of the delay increases with U. The velocity in the jet after the swell of jet has fully swelled is subcritical $U_f<U$ where U_f is the final U. This is something like a hydraulic jump with supercritical flow ahead of the delay and subcritical flow behind it.

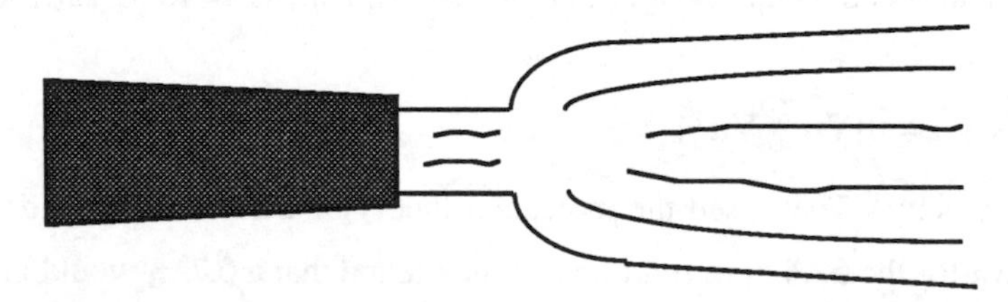

Figure 2. Delayed die swell.

Yoo and Joseph [1985] studied Poiseuille flow of an upper convected Maxwell model through a plane channel. Ahrens, Yoo and Joseph [1987] studied the same problem in a round pipe. In both cases, we get a hyperbolic region of flow in the center of the pipe when the centerline velocity U_m, equal to 2U in the Maxwell model, is greater than the wave speed c. This gives theoretical support to the idea that delayed die swell is a supercritical phenomenon.

There is a marked difference between the shape of the swell when it is delayed between different polymer solutions. The shape seems to correlate with a relaxation time

$$\lambda = \tilde{\mu}/G_c \tag{2}$$

where $\tilde{\mu}$ is the zero shear rate viscosity and G_c is the rigidity. We get G_c from measuring c

$$c^2 = G_c/\rho\ . \tag{3}$$

When λ is large, say $\lambda \geq 0(10^{-3}\text{ sec})$, the delay is sharp, as in figure 2. When the relaxation times are small, $\lambda \leq 0(10^{-4}\text{ sec})$, the delay is smoothed; in the extreme cases it is difficult to see that the swell is actually delayed.

We can say the Newtonian fluids are fluids with very large values of λ. In the case of delayed die swell, the smoothing of the swell is probably associated by the effect of smoothing due to an effective viscosity which arises from rapidly relaxing modes which have already relaxed when the delayed swell commences. Very viscous liquids always exhibit relaxation or non-Newtonian effects because even though the relaxation is fast, there is so much to relax.

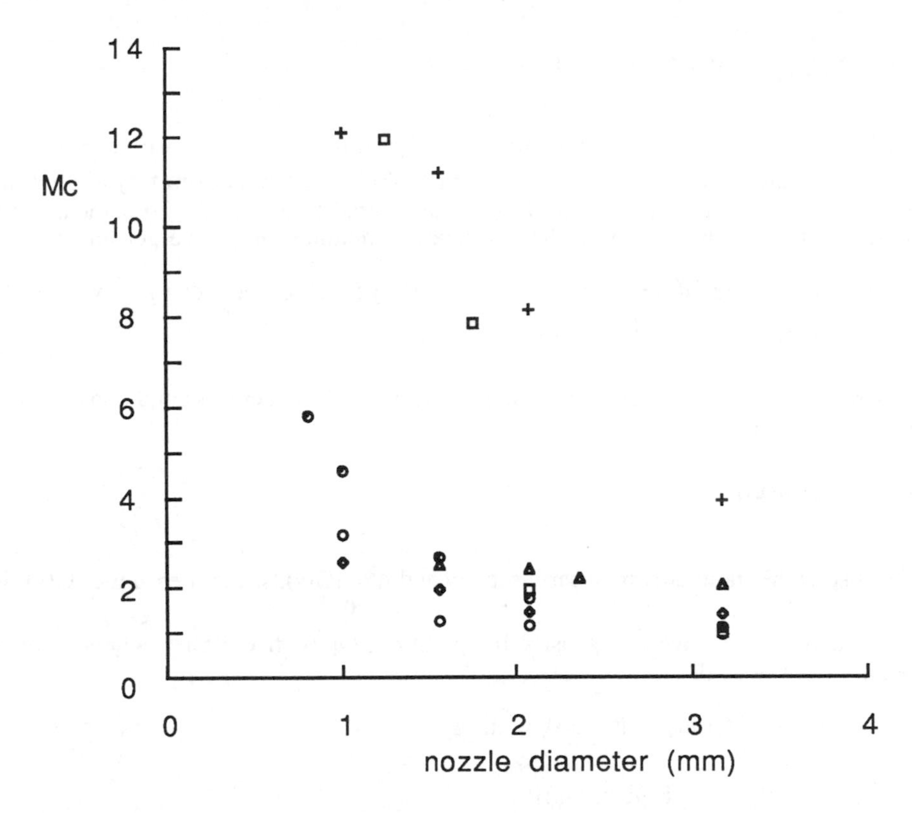

Figure 3. Mach number vs. pipe diameter.

○ 1.3% CMC □ 9.8% ELVACITE ◇ 6% PIB/D

● 2.5% POLYOX + 12.1% K-125 △ M-1

In figure 3, we plotted the critical Mach number

$$M_c = 2U_c/c$$

against the diameter of the pipe. In all cases $M_c \geq 1$, nearly. The value $M_c=1$ seems to be some form of asymptote for large values of the pipe diameter d. We do not understand why different fluids have such different M_c vs. d curves. We have thought about the consequences of shear thinning, which are important for some of the test liquids, in trying to collapse the experimental curves for different liquids into one curve, but we have not been successful.

2. Conceptual ideas

Nonlinear constitutive modeling is a jungle. The possible responses of the material to stresses are too complicated to describe by one explicit expression. General expressions are too abstract to be of direct use and are always insufficiently general to describe everything. Linearizing around rest is good because many different models collapse to one. The nonlinear parameters go away. Moreover, the elasticity of liquids is preeminently associated with propagation of small amplitude waves into rest.

We start with Boltzmann's expression for the extra stress τ which has been generalized to contain a Newtonian term

$$\tau = 2\mu\mathbf{D}\,[u(\mathbf{x}, t)] + 2\int_0^\infty G(s)\mathbf{D}[\mathbf{u}(\mathbf{x}, t-s)]ds \tag{4}$$

where $\mathbf{u}$ is the velocity, $\mathbf{D}$ is the symmetric part of grad $\mathbf{u}$ and G(s) is positive, bounded and monotonically decreasing to zero. The actual stress $\mathbf{T}=-p\mathbf{1}+\tau$ differs from τ by a "pressure" p. Equation (4) is the most general linear functional of grad $\mathbf{u}$ in a fluid. To name a fluid, we need a Newtonian viscosity μ and a shear relaxation modulus G(s). We get Jeffreys' model from (4) when we write $G(s)=\frac{\eta}{\lambda}\exp(-s/\lambda)$ and Jeffreys' model reduces to Maxwell's if also $\mu=0$.

Now we consider viscosity. In steady flow, $\mathbf{u}$ is independent of t and comes out of the integral in (4). We get

$$\tau = 2\tilde{\mu}\mathbf{D}[\mathbf{u}(\mathbf{x})] \tag{5}$$

where $\tilde{\mu}=\mu+\eta$ is the static or zero shear viscosity and $\eta = \int_0^\infty G(s)ds$, the area under G(s), is the elastic viscosity. We have a viscosity inequality $\tilde{\mu}\geq\eta$ with equality when there is no Newtonian viscosity $\mu=0$.

Now we consider elasticity $\mu=0$, writing

$$\mathbf{D}[\mathbf{u}(\mathbf{x}, t-s)] = -\frac{\partial}{\partial s}\mathbf{E}\,[\xi(\mathbf{x}, t-s)] \tag{6}$$

where ξ is a displacement and $\mathbf{E}$ is the infinitesimal strain. If it were possible to make a step in strain without flow, and it isn't possible, we would have $\mathbf{D}[\mathbf{u}(\mathbf{x}, t)] = \mathbf{E}_0(x)\delta(t)$ for Dirac δ. Then, from (4), with $\mu=0$,

$$\tau = 2G(t)\,\mathbf{E}_0(\mathbf{x})$$

and you can see why G(t) is called the stress relaxation function and G(0) the rigidity or shear modulus. Another way to see elasticity with $\mu=0$ is to write

$$\tau = 2\int_0^\infty -\frac{\partial}{\partial s}\{G(s)\mathbf{E}\,[\xi\,(\mathbf{x}, t-s)]\}ds + 2\int_0^\infty G'(s)\mathbf{E}\,[\xi\,(\mathbf{x}, t-s)]\,ds\,. \tag{7}$$

Now we can suppose that G(s) decays ever so slowly so that the second integral will tend to zero while the first gives rise to linear elasticity for an incompressible solid

$$\tau = 2G(0)\,\mathbf{E}\,[(\xi\,(\mathbf{x}, t)]. \tag{8}$$

Now we restore the Newtonian viscosity and we note that this viscosity smooths discontinuities. For example, in the problem of the suddenly accelerated plate, the boundary at y=0 below a semi-infinite plate is suddenly put into motion, sliding parallel to itself with a uniform speed. If $\mu=0$, this problem is governed by a telegraph equation. The news of the change in the boundary value from zero to constant velocity propagates into the interior by a damped wave with a velocity $c=\sqrt{G(0)/\rho}$. The amplitude of the velocity shock decays exponentially. A short while after the wave passes, the solution at the given y looks diffusive. If $\mu\neq 0$, and is small, a sharp front cannot propagate. Instead we get a shock layer whose thickness is proportional to $\sqrt{\mu y/\tilde{\mu}}$ and the solution, as in the Newtonian fluid, is felt instantly everywhere. We get a diffusive signal plus a wave. The wave could be dominant in the dynamics if μ is small.

Actually diffusion is impossible because it requires that a pulse initiated at any point be felt instantly everywhere. This same defect hold for all models with $\mu\neq 0$, like Jeffreys'. Propagation should proceed as waves.

Poisson, Maxwell, Poynting and others thought that $\mu=0$ ultimately. It's all a matter of time scales. Short range forces between molecules of a liquid give rise to weak clusters of molecules which resist fast deformations elastically, then relax. Liquids are closer to solids than to gases. Liquid molecules do not bounce around with a mean free path, they move cooperatively.

So what is the difference between two liquids with the same η, one appearing viscous (Newtonian) and the other elastic? Maxwell thought that viscous liquids were actually elastic, with high rigidity and a single fast time of relaxation. To fix his idea in your mind, we compare two liquids with the same viscosity η, satisfying Maxwell's model with $G(s)=G(0)\exp(-s/\lambda)$, $G(0)=\eta/\lambda$. To have the same η the Newtonian liquid would have a relatively large $G(0)$ and a small time λ of relaxation. The trouble with Maxwell's model, if not his idea, is that a single time of relaxation is against experiments which can never be made to fit a single time of relaxation.

There are many different times of relaxation. Experiments indicate that many liquids respond to high frequency ultrasound like a solid organic glass with

$$G(0) \sim 10^9\text{Pa}, \quad c=\sqrt{G(0)/\rho} \sim 10^5 \text{ cm/sec}. \tag{9}$$

This type of estimation is valid for a huge range of liquids, from olive oil to high molecular weight silicon oils. With this time of relaxation and such a high rigidity, all the liquids would look Newtonian, with t much greater than $\tilde{\mu}/G(0)$, which is of the order of 10^{-10} sec. in olive oil, and is perhaps 10^{-6} in some high viscosity silicon oils. In fact, we see much longer lasting responses which come about because there are different times of relaxation. Small molecules relax rapidly, giving rise to large rigidity $G(0)$ and fast speed. Large molecules and polymers relax slowly, giving rise to a smaller effective rigidity $G_\mu(0)$, effective viscosity μ and slow speed

$$c=c_\mu=\sqrt{G_\mu(0)/\rho}\,. \tag{10}$$

To get this firmly in mind, we can think of a kernel with values like those given by (9), sketched in figure 4.

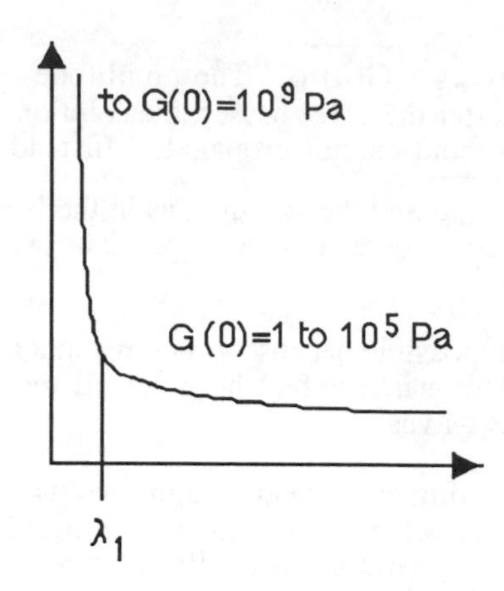

Figure 4. G(s), fast relaxation (say 10^{-10} sec) followed by a slow relaxation (say 10^{-4} sec).

We may inquire if at $t \gg 10^{-10}$ sec the relaxed fast modes have a dynamical effect. Yes, they give rise to an effective viscosity. We may as well collapse the glassy mode into a one-sided delta function $\mu\delta(s)$ where $\mu=G(0)\lambda_1$, or some fraction of this. This is our effective viscosity and our construction shows that is not unique. This is a very interesting concept, but it is not amenable to experiments that we know.

It is useful to define a time unit in terms of the slowest relaxation, say $\tilde{\mu}/G_c$. This gives rise to an internal clock, with a material time defined by the slowest relaxation. This time may be slow or fast on the external clock. To get this idea, think of the analog for the transport of heat. Heat is transported in solids by fast waves. The fastest wave is associated with electrons with relaxation times of 10^{-13} sec, then by lattice waves (phonons) with relaxation times of 10^{-11} sec. Both times are surpassingly short on our clock. However, at 10^{-13} sec, the electrons have all relaxed (and they give rise to diffusion) whilst the phonons have not begun to relax. Of course, it's more interesting when the slow relaxation is not too fast on our clock, as is true for viscoelastic fluids.

The notion of an external and internal clock is an appealing idea for expressing the difference between different theories of fading memory. Some theories, like Maxwell's and the more mathematical one by Coleman and Noll [1960] use an external clock; in rapid deformations the fluid responds elastically; in slow deformations the response is viscous. Fast and slow are measured in our time, on the external clock. Such theories rule out transient Newtonian responses. Models with $\mu\neq0$, like Jeffreys', or the more mathematical one by Saut and Joseph [1983], are disallowed. To get $\mu\neq0$ back in, even though ultimately $\mu=0$, we need an effective μ, associated with an internal clock.

3. Mathematical theory

When the fluid is elastic the governing equations are partly hyperbolic. The hyperbolic theory makes sense when the Newtonian viscosity is zero or small relative to the static viscosity $\tilde{\mu}$. For very fast deformations in which the fluid responds momentarily like a glass, the equations always exhibit properties of hyperbolic response, waves and change of type. However, the glassy response takes place in times too short to notice. Hence, the hyperbolic theory is not useful where it is exact. The hyperbolic theory is useful when we get an elastic response at times we read on our clock, in the domain of the effective theory. Hence, the hyperbolic theory is useful where it is not exact.

Most of the mathematical work has been done with fluids like Maxwell's and for plane flows. These problems are governed by six quasilinear equations in six unknowns. The unknowns are two velocity components, three components of the stress, and a pressure. The continuity equation, two momentum equations and three equations for the stress govern the evolution of the six variables. The stress equations are like Maxwell's

$$\lambda\left[\frac{\partial\tau}{\partial t}+\mathbf{u}\cdot\nabla\tau+\tau\Omega-\Omega\tau-a(\mathbf{D}\tau+\tau\mathbf{D})\right]=2\eta\mathbf{D}+\ell$$

where $\mathbf{D}$ is the symmetric part and Ω the antisymmetric part of $\nabla\mathbf{u}$, $-1\leq a\leq 1$ and ℓ are lower order terms, algebraic in the system variables. This system may be analyzed for type in the usual way. We get a 6th order system and it factors into three quadratic roots. Two of the roots are imaginary so that the system is not hyperbolic. The streamlines are characteristic, with double roots so that the system is not strictly hyperbolic. The third quadratic factor depends on the unknown solution, algebraically, and it can be real or complex, depending on the solution. We say that such a solution with mixed roots is of composite type. Some variables are elliptic, some are hyperbolic.

It turns out that the pair of roots which depend on the unknown solution and can change type are associated with the vorticity equation, a second order nonlinear PDE. This equation is either elliptic or it is hyperbolic, depending on the solution. It is not of composite type, but is classical, like the equation for the potential in gas dynamics.

We can think of the unsteady vorticity equation and the steady vorticity equation. The analysis of the two has greatly different consequences. The unsteady equation is ill-posed when it is elliptic and well-posed when it is hyperbolic. Ill-posed problems are catastrophically unstable to short waves, with growth rates which go to infinity with the wave number. The conditions on the stress which lead to ill-posed problems can be determined by the method of frozen coefficients, as was first done by Rutkevich [1969]. It turns out that the Maxwell models with $a=\pm 1$ cannot be ill-posed on smooth solutions, but the other models do become ill-posed for certain flows.

The problem of change of type in steady flow is different. The vorticity in steady flow can be of mixed type with elliptic and hyperbolic regions, as in transonic flow. The physical implications of these mixed "transonic" fields are not yet perfectly understood, though many examples have been calculated.

There are many models, other than those like Maxwell's, in which vorticity is the key variable. It is the only variable which is either strictly elliptic or strictly hyperbolic. The stream function satisfies Laplace's equation, the velocity and the stresses are of composite type. The stresses do not satisfy a hyperbolic equation and it is wrong to speak of the propagation of stress waves.

There are other models in which the vorticity is not the key variable. However, when these models are linearized around rest, one finds again that the steady vorticity equation is either elliptic or hyperbolic, and the unsteady vorticity equation is always hyperbolic. Hence it is precisely waves of vorticity which propagate into rest.

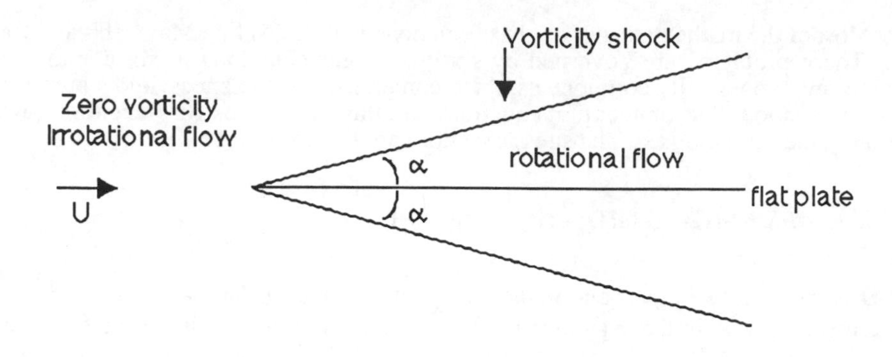

Figure 5. Mach wedge for the vorticity, tan $\alpha = (1-M^2)^{-1/2}$.

The mathematical consequences of composite roots are clearly evident in the recent solution of L. E. Fraenkel [1987] of the problem of linearized supercritical flow over a flat plate. The linearization here is around the uniform flow which exists at infinity, as in Oseen's problem for the Navier-Stokes equation. Fraenkel's solution shows that there is a Mach wedge of vorticity ζ centered on the leading edge of the plate. The vorticity in front of this wedge is zero and it is not zero behind the wedge [see figure 5]. Surprisingly, the vorticity jumps from zero to infinity at the wedge, but the singularity is integrable. We have rotational flow behind the shock and irrotational flow in front of the shock. The stream function satisfies $\nabla^2\Psi=-\zeta$ where $\zeta=0$ in front of the shock. Therefore, we may write $\Psi=\Psi_1+\Psi_2$, $\nabla^2\Psi_2=-\zeta$, $\nabla^2\Psi_1=0$. To satisfy the boundary conditions on the plate, we must have a nonzero potential field Ψ_1. In fact Ψ_1 satisfies a Dirichlet problem for the region outside a strip on the positive x axis.

The potential flow decays to uniform flow as one moves upstream, but the delay is slow. There is no upstream influence in the fully hyperbolic flow of a gas over a flat plate. The upstream influence of the flat plate in the flow of a Newtonian fluid is almost negligible. The persistence of Ψ_1 is a consequence of its ellipticity, ultimately to the fact that the first order system is of composite type. This type of solution may be new in mathematical physics.

The velocity and the stresses decompose into harmonic and vortical parts. Hence these fields are all of composite type. Only the vorticity is pure, strictly hyperbolic in the linearized problem of flow past bodies. The velocity and stresses are continuous across the shock. The normal derivative of the velocity, the normal and shear stress are also continuous, but the tangential derivative of the tangential components of velocity and stress are discontinuous. The elliptic component of our composite system is associated with a huge upstream influence.

As a final matter I should like to discuss the question of the blow up of certain variables in unsteady flow of nonlinear models of a viscoelastic fluid. For one-dimensional shearing flows this question reduces to whether the vorticity or the velocity blows up as a consequence of nonlinear evolution starting from smooth data. The general belief, without proof, is that the blow up of vorticity will lead to a shock of velocity and the blow up of velocity to a shock in an integral of the velocity, say the stream function. I am going to develop my ideas in a rather more general context, calling attention to the difference between first order quasilinear systems in which derivative $\partial \mathbf{u}/\partial x_i$ of a system variable $\mathbf{u}$ appears linearly and a nonlinear first order system in which the first derivatives appear in a nonlinear way, say $\frac{\partial u_1}{\partial x}\frac{\partial u_2}{\partial y}$.

We can reduce a system of N first order nonlinear partial differential equations in γ independent variable to a system of $(\gamma+1)N$ quasilinear equations. A general system of N first order PDE's in two independent variables can be expressed as

$$F_i\,(x, y, u_1, \ldots, u_n, p_1, \ldots, p_n, q_1, \ldots, q_n) = 0 \qquad i = 1, \ldots, n \tag{12}$$

where

$$p_i = \frac{\partial u_i}{\partial x} \qquad q = \frac{\partial u_i}{\partial y}$$

are introduced as additional unknowns. We have 3N variables and 3N equations but one of the equations is nonlinear rather than quasilinear. The system can be reduced to a quasilinear one by differentiation but the reduction is not unique. One symmetric reduction is: $F_i = 0$ is an identity in x and y jointly, hence

$$\frac{dF_i}{dy} = \frac{\partial F_i}{\partial y} + \frac{\partial F_i}{\partial u_\ell} q_\ell + \frac{\partial F_i}{\partial q_\ell}\frac{\partial q_\ell}{\partial y} + \frac{\partial F_i}{\partial p_\ell}\frac{\partial p_\ell}{\partial y} = 0\,, \tag{13}$$

$$\frac{dF_i}{dx} = \frac{\partial F_i}{\partial x} + \frac{\partial F_i}{\partial u_\ell} p_\ell + \frac{\partial F_i}{\partial q_\ell}\frac{\partial q_\ell}{\partial x} + \frac{\partial F_i}{\partial p_\ell}\frac{\partial p_\ell}{\partial x} = 0\,, \tag{14}$$

Equation (12) implies that

$$\frac{\partial F_i}{\partial p_j} p_j + \frac{\partial F_i}{\partial q_j} q_j = \frac{\partial F_i}{\partial p_j}\frac{\partial u_j}{\partial x} + \frac{\partial F_i}{\partial q_j}\frac{\partial u_j}{\partial y} \tag{15}$$

We put this system into a symmetric form by writing $\partial p_\ell/dy = \partial q_\ell/dx$ in (14) and (15). Then we put the principal part on the right and the lower order terms on the left

$$-\left(\frac{\partial F_i}{\partial y} + \frac{\partial F_i}{\partial u_j} q_j\right) = \frac{\partial F_i}{\partial p_j}\frac{\partial q_i}{\partial x} + \frac{\partial F_i}{\partial q_j}\frac{\partial q_i}{\partial y} \tag{16}$$

$$-\left(\frac{\partial F_i}{\partial x} + \frac{\partial F_i}{\partial u_j} p_j\right) = \frac{\partial F_i}{\partial p_j}\frac{\partial p_i}{\partial x} + \frac{\partial F_i}{\partial q_j}\frac{\partial p_i}{\partial y} \tag{17}$$

Equations (15), (16), and (17) are 3N equations for the 3N unknowns.

The principal parts of each of the equations (15), (16), and (17) are identical. Each one determines the same characteristic directions. We have

$$\frac{\partial F_i}{\partial p_j}\frac{\partial a_j}{\partial x} + \frac{\partial F_i}{\partial q_j}\frac{\partial a_j}{\partial y} = \text{lot}\,. \tag{18}$$

Hence the characteristics $\lambda = dy/dx$ are determined from

$$\det\left[\lambda\,\frac{\partial F_i}{\partial p_j} - \frac{\partial F_i}{\partial q_j}\right] = 0\,. \tag{19}$$

Equation (19) has N roots. The N nonlinear first order PDE's give rise to N characteristic roots for the quasilinear system arising from differentiating the nonlinear system once with respect to each independent variable.

If we generate (19) by the method of simple jumps, we can state that real characteristic directions are the loci for discontinuities in the derivatives of p_i and q_i. This means that the second derivatives of u_i suffer jumps in the nonlinear case and first derivatives jump in the quasilinear case. The first derivatives are smooth when second derivatives jump so that we get one more derivative of smoothness in the nonlinear case.

It appears that a more far-reaching conclusion following along lines of the last paragraph may be true. Compare quasilinear and nonlinear first order systems which allow blow up in finite time. The solutions are smooth before the blow up time. To find blow up we look for intersecting characteristics. First derivatives blow up in quasilinear systems, second derivatives in nonlinear systems. This conjecture is true for some special one-dimensional models of flow of a viscoelastic fluid which have been studied by M. Slemrod [1985] and Renardy, Hrusa, and Nohel [1987].

Acknowledgement

This work was supported by the Army Research Office, National Science Foundation and Department of Energy.

References

Ahrens, M., Yoo, J. Y. and Joseph, D. D. 1987 Hyperbolicity and change of type in the flow of viscoelastic fluids through pipes, *J. Non-Newtonian Fluid Mech.* **24**, 67–83.

Ambari, A., Deslouis, C. and Tribollet, B. 1984 Coil-stretch transition of macromolecules in laminar flow around a small cylinder. *Chem. Eng. Commun.* **29**, 63–78.

Coleman, B. and Noll, W. 1960 An approximation theorem for functionals with applications in continuum mechanics, *Arch. Rat. Mech. Anal.* **6**, 355–370.

Fraenkel, L. E. 1987 Some results for a linear, partly hyperbolic model of viscoelastic flow past a plate, in *Material Instabilities in Continuum Mechanics and Related Mathematical Problems* (ed. J. M. Ball) Clarendon Press: Oxford.

James, D. and Acosta, A. J. 1970 The laminar flow of dilute polymer solutions around circular cylinders, *J. Fluid Mech.* **42**, 269–288.

Joseph, D. D., Matta, J. and Chen, K. 1987 Delayed die swell, *J. Non-Newtonian Fluid Mech.* **24**, 31–65.

Konuita, A., Adler, P. M. and Piau, J. M. 1980 Flow of dilute solutions around circular cylinders. *J. Non-Newtonian Fluid Mech.* **7**, 101–106.

Renardy, M., Hrusa, W. J. and Nohel, J. A. 1987 *Mathematical Problems in Viscoelasticity.* Longman: Harlow, UK.

Rutkevich, M. I. 1969 Some general properties of the equations of viscoelastic fluid dynamics. *PMM* **33**, 42–51.

Saut, J. C. and Joseph, D. D. 1983 Fading memory, *Arch. Rat. Mech. Anal.* **81**, 53–95.

Slemrod, M. 1985 Breakdown of smooth shearing flow in viscoelastic fluids for two constitutive equations; the vortex sheet vs. the vortex shock. Appendix to the paper by Joseph, D. D. Hyperbolic phenomena in the flow of viscoelastic fluids, in *Viscoelasticity and Rheology* (eds. A. S. Lodge, J. Nohel and M. Renardy), Academic Press.

Ultman, J. S. and Denn, M. M. 1970 Anomalous heat transfer and a wave phenomenon in dilute polymer solutions. *Trans. Soc. Rheol.* **14**, 307–317.

Yoo, J. and Joseph, D. D. 1985 Hyperbolicity and change of type in the flow of viscoelastic fluids through channels. *J. Non-Newtonian Fluid Mech.* **19**, 15–41.

For additional references see the Appendix

ANALYSIS OF SPURT PHENOMENA FOR A NON-NEWTONIAN FLUID *

David S. Malkus [1]
John A. Nohel [2]
Bradley J. Plohr [3]

Center for the Mathematical Sciences
University of Wisconsin–Madison
Madison, WI 53705

1. Introduction

The purpose of this paper is to analyze novel phenomena in dynamic shearing flows of non-Newtonian fluids that are important for polymer processing. Understanding such behavior has proved to be of significant physical, mathematical, and computational interest. One striking phenomenon, called "spurt," was observed by Vinogradov *et al.* [20] in the flow of monodispersive polyisoprenes through capillaries. They found that the volumetric flow rate increased dramatically at a critical stress that was independent of molecular weight. Until recently, spurt had been overlooked or dismissed by rheologists because no plausible mechanism was known to explain it in the context of steady flows that are linearly stable.

We find that satisfactory explanation and modeling of the spurt phenomenon requires studying the full dynamics of the equations of motion and constitutive equations. The common feature of constitutive models that exhibit spurt is a non-monotonic relation between the steady shear stress and strain rate. This allows jumps in the steady strain rate to form when the driving pressure gradient exceeds a critical value; such jumps correspond to the sudden increase in volumetric flow rate observed in the experiments of Vinogradov *et al.* Hunter and Slemrod [7] studied the qualitative behavior of these jumps in a one-dimensional viscoelastic model of rate type and predicted shape memory and hysteresis effects related to spurt. A salient feature of this model is linear instability and loss of evolutionarity in a region of state space. By contrast, the equations that are analyzed in the present work derive from a fully three-dimensional constitutive relation and remain stable and evolutionary, as we would expect of a realistic model. This model also exhibits spurt, shape memory, and hysteresis; furthermore, it predicts other effects, such as latency, normal stress oscillations, and molecular weight dependence of hysteresis, that can be tested in rheological experiment.

In Refs. [9, 12, 13], effective numerical methods were developed for simulating one-dimensional shear flows at high Weissenberg (Deborah) number; calculations using these methods agreed qualitatively and quantitatively with experiment. We discussed

---

* Supported by the U. S. Army Research Office under Grant DAAL03-87-K-0036, the National Science Foundation under Grants DMS-8712058 and DMS-8620303, and the Air Force Office of Scientific Research under Grants AFOSR-87-0191 and AFOSR-85-0141.

[1] Also Department of Engineering Mechanics.

[2] Also Department of Mathematics.

[3] Also Computer Sciences Department.

preliminary results on global existence and stability of discontinuous steady states, and we introduced a system of ordinary differential equations that approximate the dynamics of highly elastic and very viscous fluids, such as those in the experiments of Vinogradov *et al.* The objective of the present paper is to analyze this approximating dynamical system. Based on this analysis, we explain the shape memory, hysteresis, latency, and other effects that have been observed in the numerical simulations. Similar results are known for related constitutive models: a model with two or more relaxation times and no Newtonian viscosity is analyzed mathematically in Ref. [14]; and another model with a single relaxation time and Newtonian viscosity is studied numerically in Ref. [10]. Therefore we believe that our results are not limited to the specific model that we study.

The paper is organized as follows: Sec. 2 formulates and discusses the flow model; Sec. 3 provides a complete description of the dynamics of the approximating quadratic system of ordinary differential equations by means of a phase plane analysis; Sec. 4 uses the results of Sec. 3 to describe features of the mathematical model in relation to the experiments of Vinogradov *et al.* and explains latency, shape memory, and hysteresis analytically; and Sec. 5 discusses certain physical and mathematical conclusions.

2. The Flow Model

The motion of a fluid under incompressible and isothermal conditions is governed by the balance of mass and linear momentum. The response characteristics of the fluid are embodied in the constitutive relation for the stress. For viscoelastic fluids with fading memory, these relations specify the stress as a functional of the deformation history of the fluid. Many sophisticated constitutive models have been devised; see Ref. [1] for a survey. In the present work, we focus on a particular differential model that is explained in more detail in Ref. [13]. This model can be regarded as a special case of the Johnson-Segalman model [8] and of the Oldroyd constitutive equation [17]. We believe, however, that qualitative aspects of our results are not limited to this particular model; results on similar models [14, 10] confirm this.

Essential properties of constitutive relations are exhibited in simple planar Poiseuille shear flow. We study the Poiseuille shear flow between parallel plates located at $x = \pm h/2$, with the flow aligned along the y-axis (see Fig. 1). Therefore, the flow variables are independent of y, and the velocity field is $\mathbf{v} = (0, v(x,t))$, which implies that the balance of mass is automatically satisfied. The stress is decomposed into three parts: an isotropic pressure p; a Newtonian contribution, characterized by viscosity η; and an extra stress, characterized by a shear modulus μ and a relaxation rate λ. In shear flow, the components of the extra stress tensor Σ can be written $\Sigma^{xx} = Z(x,t)/(1+a)$, $\Sigma^{xy} = \Sigma^{yx} = \sigma(x,t)$, and $\Sigma^{yy} = -Z(x,t)/(1-a)$, while the pressure takes the form $p = p_0(x,t) - f(t)y$, f being the pressure gradient driving the flow. Here $a \in (-1,1)$ is a slip parameter defining the model.

To simplify notation, we nondimensionalize the variables by scaling distance by h, time by λ^{-1}, and stress by μ. Furthermore, if we replace σ, v, and f by $\hat{\sigma} := (1-a^2)^{1/2}\sigma$, $\hat{v} := (1-a^2)^{1/2}v$, and $\hat{f} := (1-a^2)^{1/2}f$, respectively, then the parameter a disappears

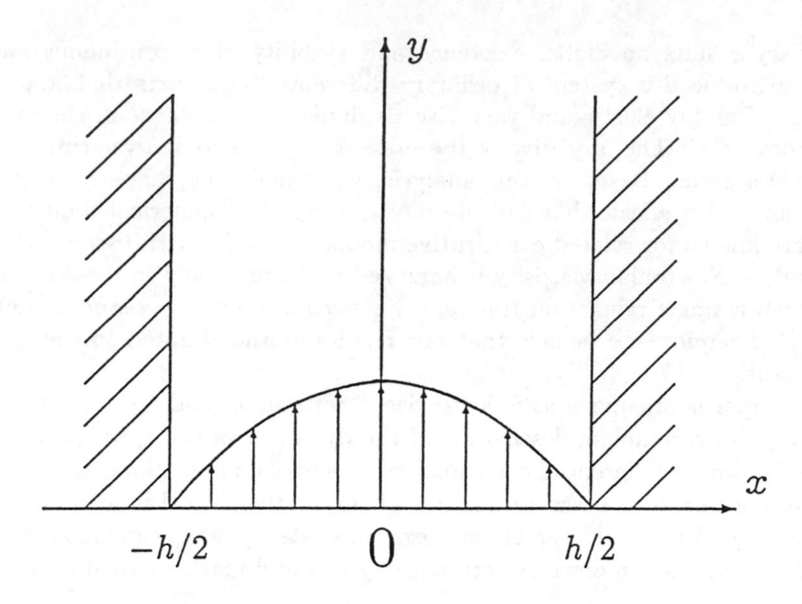

Fig. 1: Shear flow through a slit-die.

from the governing equations. Since no confusion will arise, we omit the caret. There are two essential dimensionless parameters:

$$\alpha := \rho h^2 \lambda^2 / \mu \ , \tag{2.1}$$

a ratio of Reynolds number to Deborah number (ρ being the constant density); and

$$\varepsilon := \eta \lambda / \mu \ , \tag{2.2}$$

a ratio of viscosities.

The resulting initial–boundary-value problem governing the flow [13] is the system

$$\begin{aligned} \alpha v_t - \sigma_x &= \varepsilon v_{xx} + f \ , \\ \sigma_t - (Z+1) v_x &= -\sigma \ , \\ Z_t + \sigma v_x &= -Z \end{aligned} \tag{JS}$$

on the interval $[-1/2, 0]$, with boundary conditions

$$v(-1/2, t) = 0 \quad \text{and} \quad v_x(0, t) = 0 \tag{BC}$$

and initial conditions

$$v(x,0) = v_0(x) , \quad \sigma(x,0) = \sigma_0(x) , \quad \text{and} \quad Z(x,0) = Z_0(x) , \qquad (IC)$$

where the compatibility conditions $v_0(-1/2) = 0$, $v_0'(0) = 0$ and $\sigma_0(0) = 0$ are assumed to hold.

If $\varepsilon = 0$, the system can be classified according to type: when $Z + 1 \geq 0$, the system is hyperbolic, with characteristics speeds 0 and $\pm[(Z+1)/\alpha]^{1/2}$; if $Z + 1 < 0$, by contrast, the system has a pair of pure imaginary characteristic speeds, in addition to the speed 0, and (JS) ceases to be evolutionary. This classification, however, is not applicable when $\varepsilon > 0$, which we assume throughout the present work. In the case $\varepsilon > 0$, it was shown recently [6] that the problem (JS), (BV), (IC) possesses a unique classical solution globally in time for smooth initial data of arbitrary size. However, it is has not been proved that the solution tends to a limiting steady state as t tends to infinity.

The steady-state solution of system (JS), when the forcing term f is a constant $\overline{f}$, plays an important role in our discussion. Such a solution, denoted by $\overline{v}$, $\overline{\sigma}$, and $\overline{Z}$, can be described as follows. The stress components $\overline{\sigma}$ and $\overline{Z}$ are related to the strain rate $\overline{v}_x$ through

$$\overline{\sigma} = \frac{\overline{v}_x}{1 + \overline{v}_x^2} \tag{2.3}$$

and

$$\overline{Z} + 1 = \frac{1}{1 + \overline{v}_x^2} \; . \tag{2.4}$$

Therefore, the steady total shear stress $\overline{T} := \overline{\sigma} + \varepsilon \overline{v}_x$ is given by $\overline{T} = w(\overline{v}_x)$, where

$$w(s) := \frac{s}{1 + s^2} + \varepsilon s \; . \tag{2.5}$$

The properties of w, the steady-state relation between shear stress and shear strain rate, are crucial to the behavior of the flow. By symmetry, it suffices to consider nonnegative strain rates, $s \geq 0$. For all $\varepsilon > 0$, the function w has inflection points at $s = 0$ and $s = \sqrt{3}$. When $\varepsilon > 1/8$, the function w is strictly increasing, but when $\varepsilon < 1/8$, the function w is not monotone. Lack of monotonicity is the fundamental cause of the non-Newtonian behavior studied in this paper, so hereafter we assume that $\varepsilon < 1/8$.

The graph of w is shown in Fig. 2. Specifically, w has a maximum at $s = s_M$ and a minimum at $s = s_m$, where

$$s_M, s_m = \left[\frac{1 - 2\varepsilon \mp \sqrt{1 - 8\varepsilon}}{2\varepsilon} \right]^{1/2} , \tag{2.6}$$

respectively, at which points it takes the values $\overline{T}_M := w(s_M)$ and $\overline{T}_m := w(s_m)$. As $\varepsilon \to 1/8$, the two critical points coalesce at $s = \sqrt{3}$.

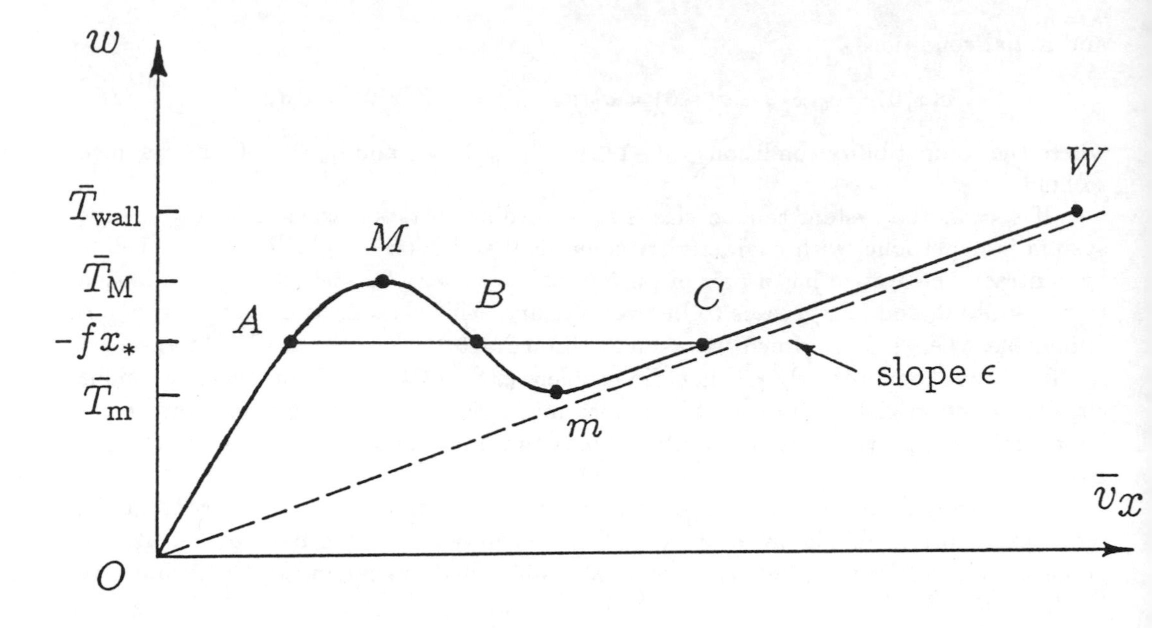

Fig. 2: Total steady shear stress $\overline{T}$ *vs.* shear strain rate $\overline{v}_x$ for steady flow. The case when $\overline{f} > 0$ and there are three critical points is illustrated; other possibilities are discussed in Secs. 3 and 4.

The momentum equation, together with the boundary condition at the centerline, implies that the steady total shear stress satisfies $\overline{T} = -\overline{f}x$ for every $x \in [-\frac{1}{2}, 0]$. Therefore, the steady velocity gradient can be determined as a function of x by solving

$$w(\overline{v}_x) = -\overline{f}x \ . \tag{2.7}$$

Equivalently, a steady state solution $\overline{v}_x$ satisfies the cubic equation $P(\overline{v}_x) = 0$, where

$$P(s) := \varepsilon\, s^3 - \overline{T}\, s^2 + (1+\varepsilon)s - \overline{T} \ . \tag{2.8}$$

The steady velocity profile in Fig. 3 is obtained by integrating $\overline{v}_x$ and using the boundary condition at the wall. However, because the function w is not monotone, there might be up to three distinct values of $\overline{v}_x$ that satisfy Eq. (2.7) for any particular x on the interval $[-1/2, 0]$. Consequently, $\overline{v}_x$ can suffer jump discontinuities, resulting in kinks in the velocity profile (as at the point x_* in Fig. 3). Indeed, a steady solution must contain such a jump if the total stress $\overline{T}_{\text{wall}} = |\overline{f}|/2$ at the wall exceeds the total stress $\overline{T}_M$ at the local maximum M in Fig. 2.

Equation (2.7) implies that the stress $\Sigma := \sigma + \varepsilon v_x + \overline{f}x$ vanishes throughout the channel in steady state. The numerical simulations of (JS) in Ref. [13] suggest that Σ

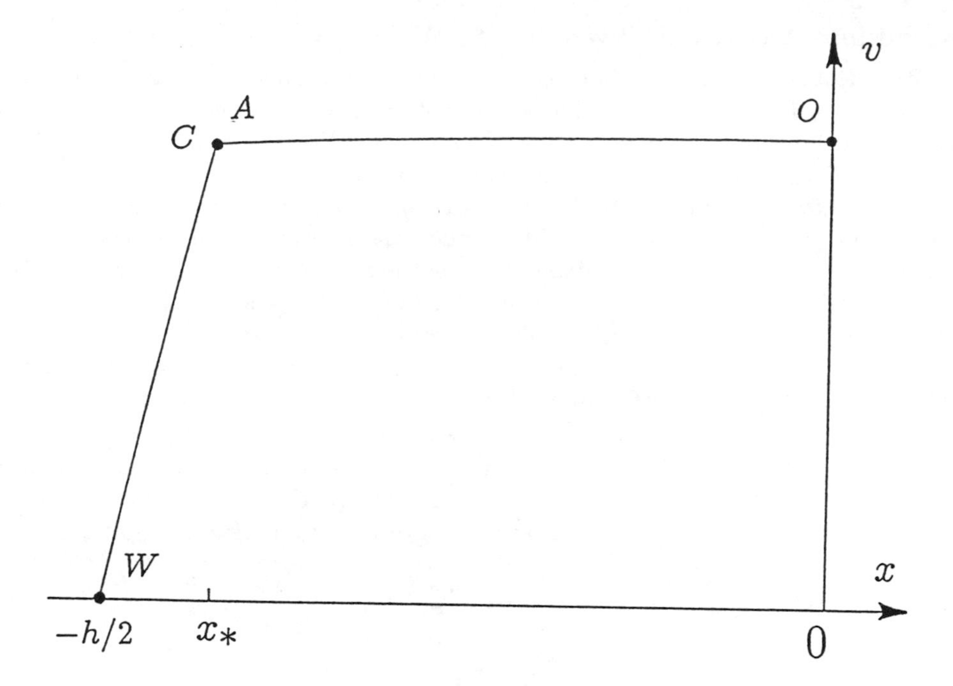

Fig. 3: Velocity profile for steady flow.

tends to zero as t tends to infinity for every x, even though the limiting solution can suffer discontinuities in $\overline{v}_x$ and $\overline{\sigma}$ whenever $-|\overline{f}|x$ lies between the values $\overline{T}_m$ and $\overline{T}_M$ of w in Fig. 2. Therefore the discontinuities in σ and εv_x seem to cancel. This behavior has been proved recently [16] for a simpler model problem that captures several key features of (JS). It is also shown that the discontinuous steady solution is stable with respect to small perturbations of initial data. Current work of Nohel, Pego, and Tzavaras indicates that a similar stability result holds for solutions of (JS) whenever the parameter α is sufficiently small.

3. Phase Plane Analysis for System (JS) When $\alpha = 0$

A great deal of information about the structure of solutions of system (JS) can be garnered by studying a system of ordinary differential equations that approximates it in a certain parameter range. Motivation for this approximation comes from the following observation: in experiments of Vinogradov *et al.* [20], $\alpha = \rho h^2 \lambda^2/\mu$ is of the order 10^{-12}; thus the term αv_t in the momentum equation of system (JS) is negligible even when v_t is moderately large. We are led to study the approximation to system (JS) obtained when $\alpha = 0$. The behavior of solutions of the resulting dynamical system offers an explanation for several features of the solutions of the full system (JS) observed in the computations of Refs. [9, 13]; in fact, these calculations prompted the following analysis, which determines the dynamics of the approximating system completely.

When $\alpha = 0$, the momentum equation in system (JS) can be integrated, just as in the case of steady flows, to show that the total shear stress $T := \sigma + \varepsilon v_x$ coincides with the steady value $\overline{T}(x) = -\overline{f}x$. Thus $T = \overline{T}(x)$ is a function of x only, even though σ and v_x are functions of both x and t. The remaining equations of system (JS) yield, for each fixed x, the autonomous planar system of ordinary differential equations

$$\begin{aligned} \dot{\sigma} &= (Z+1)\left(\frac{\overline{T}-\sigma}{\varepsilon}\right) - \sigma\ , \\ \dot{Z} &= -\sigma\left(\frac{\overline{T}-\sigma}{\varepsilon}\right) - Z\ . \end{aligned} \tag{3.1}$$

Here the dot denotes the derivative d/dt. We emphasize that a different dynamical system is obtained at each point on the interval $[-1/2, 0]$ in the channel because $\overline{T}$ depends on x. These dynamical systems can be analyzed completely by a phase-plane analysis, which we carry out in some detail.

The critical points of system (3.1) satisfy the algebraic system

$$\begin{aligned} (Z+1+\varepsilon)\left(\frac{\sigma}{\overline{T}} - 1\right) + \varepsilon &= 0\ , \\ \frac{\overline{T}^2}{\varepsilon}\frac{\sigma}{\overline{T}}\left(\frac{\sigma}{\overline{T}} - 1\right) - Z &= 0\ . \end{aligned} \tag{3.2}$$

Eliminating Z in these equations shows that the σ-coordinates of the critical points satisfy the cubic equation $Q(\sigma/\overline{T}) = 0$, where

$$Q(\xi) := \left[\frac{\overline{T}^2}{\varepsilon}\xi(\xi-1) + 1 + \varepsilon\right](\xi - 1) + \varepsilon\ . \tag{3.3}$$

Since

$$P(\overline{v}_x) = P\left(\frac{\overline{T}-\sigma}{\varepsilon}\right) = \frac{\overline{T}}{\varepsilon}Q(\sigma/\overline{T}) \tag{3.4}$$

[*cf.* Eqs. (2.8) and (3.3)], each critical point of the system (3.1) defines a steady-state solution of system (JS): such a solution corresponds to a point on the steady total-stress curve (see Fig. 2) at which the total stress is $\overline{T}(x)$.

By symmetry, we may focus attention on the case $\overline{T} > 0$. Consequently, for each position x in the channel and for each $\varepsilon > 0$, there are three possibilities:
(1) there is a single critical point A when $\overline{T} < \overline{T}_m$;
(2) there is also a single critical point C if $\overline{T} > \overline{T}_M$;
(3) there are three critical points A, B, and C when $\overline{T}_m < \overline{T} < \overline{T}_M$.
For simplicity, we ignore the degenerate cases, where $\overline{T} = \overline{T}_M$ or $\overline{T} = \overline{T}_m$, in which two critical points coalesce.

To determine the qualitative structure of the dynamical system (3.1), we first study the nature of the critical points. The behavior of orbits near a critical point depends on the linearization of Eq. (3.1) at this point, *i.e.*, on the eigenvalues of the Jacobian

$$\mathbf{J} = \begin{pmatrix} -\frac{1}{\varepsilon}(Z+1+\varepsilon) & -\frac{\overline{T}}{\varepsilon}\left(\frac{\sigma}{\overline{T}}-1\right) \\ \frac{\overline{T}}{\varepsilon}\left(2\frac{\sigma}{\overline{T}}-1\right) & -1 \end{pmatrix}, \tag{3.5}$$

evaluated at the critical point. The character of the eigenvalues of J can be determined from the signs of the trace of $\mathbf{J}$, given by

$$-\varepsilon \operatorname{Tr} \mathbf{J} = Z + 1 + 2\varepsilon \; ; \tag{3.6}$$

the determinant of $\mathbf{J}$, given by

$$\varepsilon \operatorname{Det} \mathbf{J} = Z + 1 + \varepsilon + \frac{\overline{T}^2}{\varepsilon}\left(2\frac{\sigma}{\overline{T}} - 1\right)\left(\frac{\sigma}{\overline{T}} - 1\right) \; ; \tag{3.7}$$

and the discriminant of $\mathbf{J}$, given by

$$\varepsilon^2 \operatorname{Discrm} \mathbf{J} = (Z+1)^2 - 8\overline{T}^2\left(\frac{\sigma}{\overline{T}} - \frac{3}{4}\right)^2 + \tfrac{1}{2}\overline{T}^2 \, . \tag{3.8}$$

We note a useful fact: at a critical point,

$$\varepsilon \operatorname{Det} \mathbf{J} = Q'(\sigma/\overline{T}) \; ; \tag{3.9}$$

this follows by using the second of Eqs. (3.2) to replace Z in Eq. (3.7). This relation is important because Q' is positive at A and C and negative at B.

The character of the eigenvalues can be understood using these formulae together with Fig. 4. In this figure is drawn the hyperbola on which $\dot{\sigma} = 0$ and parabola on which $\dot{Z} = 0$ [see Eqs. (3.2)]. These curves intersect at the critical points of the dynamical system for the given choice of ε and $\overline{T}$; Fig. 4 corresponds to the most comprehensive case of three critical points. Notice that, having scaled the σ-coordinate by $\overline{T}$, the hyperbola on which $\dot{\sigma} = 0$ is independent of $\overline{T}$. Also drawn in Fig. 4 is the hyperbola on which $\operatorname{Discrm} \mathbf{J}$ vanishes. We draw the following conclusions:
(1) $\operatorname{Tr} \mathbf{J} < 0$ at all critical points;
(2) $\operatorname{Det} \mathbf{J} > 0$ at A and C, while $\operatorname{Det} \mathbf{J} < 0$ at B; and
(3) $\operatorname{Discrm} \mathbf{J} > 0$ at A and B, whereas $\operatorname{Discrm} \mathbf{J}$ can be of either sign at C. (For typical values of ε and $\overline{T}$, $\operatorname{Discrm} \mathbf{J} < 0$ at C; in particular, $\operatorname{Discrm} \mathbf{J} < 0$ if C is the only critical point. But it is possible for $\operatorname{Discrm} \mathbf{J}$ to be positive if $\overline{T}$ is sufficiently close to $\overline{T}_m$.)

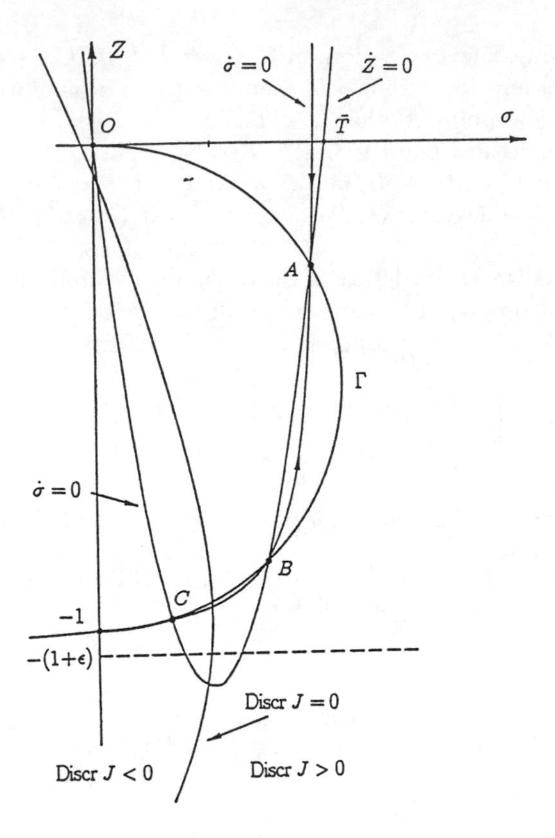

Fig. 4: The phase plane in the case of three critical points.

Standard theory of nonlinear planar dynamical systems (see, *e.g.*, Ref. [2, Chap. 15]) now establishes the local characters of the critical points:

(1) A is an attracting node (called the classical attractor);

(2) B is a saddle point; and

(3) C is either an attracting spiral point or an attracting node (called the spurt attractor).

To understand the global qualitative behavior of orbits, we construct suitable invariant sets. In this regard, a useful tool is the identity

$$\frac{d}{dt}\left\{\sigma^2+(Z+1)^2\right\} = -2\left[\sigma^2+(Z+\tfrac{1}{2})^2-\tfrac{1}{4}\right] , \tag{3.10}$$

which is obtained by multiplying the first of Eqs. (3.1) by σ and adding the second, multiplied by $Z+1$. Thus the function $V(\sigma, Z) := \sigma^2+(Z+1)^2$ serves as a Lyapunov function for the dynamical system.

Let Γ denote the circle on which the right side of Eq. (3.10) vanishes, and let C_r denote the circle of radius r centered at $\sigma = 0$ and $Z = -1$; each C_r is a level set of V. The curves Γ and C_1 are shown in Fig. 5, which corresponds to the case of a single critical point, the spiral point C; and in Fig. 7, which corresponds to the case of three critical points. Notice that if $r > 1$, Γ lies strictly inside C_r. Consequently, Eq. (3.10) shows that the dynamical system flows inward at points along C_r. Thus the interior

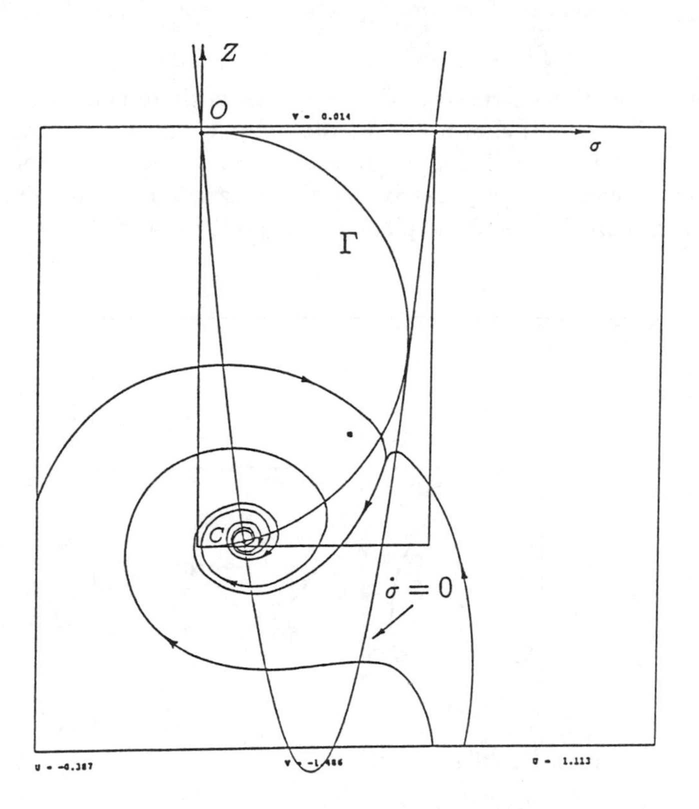

Fig. 5: The phase plane when the spurt attractor C is the only critical point.

of C_r is an invariant set for each $r > 1$. Furthermore, the closed disk bounded by C_1, which is the intersection of these sets, is also invariant. For later convenience, denote by D the point where C_1 intersects the parabola on which $\dot{Z} = 0$.

We will also rely on some theorems for quadratic dynamical systems. The hypotheses for these theorems requires an analysis of the behavior of orbits at infinity, which is accomplished as follows. First, we introduce the variables $\rho > 0$ and $\varphi \in [0, 2\pi)$ such that $\sigma = \rho^{-1}\cos\varphi$ and $Z = \rho^{-1}\sin\varphi$. Thus $\rho = 0$ defines the circle at infinity. Second, we make a singular change of independent variable, from t to s, defined by $\rho\, ds = dt$, and we let a prime denote differentiation with respect to s. Then a simple calculation shows that

$$\begin{aligned} \rho' &= \left[1 + \varepsilon^{-1}\cos^2\varphi\right]\rho^2 + O(\rho^3) \, , \\ \varphi' &= \varepsilon^{-1}\cos\varphi + O(\rho) \, . \end{aligned} \tag{3.11}$$

Therefore the critical points at infinity ($\rho = 0$) occur at the angles $\varphi = \pm\pi/2$. Correspondingly, the eigenvalues in the angular direction are $\mp\varepsilon^{-1}$, while the eigenvalues in the radial direction vanish. Because the leading order term in ρ' is positive, the critical point at $\varphi = \pi/2$ is a saddle-node for which the separatrix leaves infinity, and the critical point at $\varphi = -\pi/2$ is a repelling node.

A.

Let us first consider the structure of the flow when there is a single critical point, located at C; see Fig. 5. As shown above, the point C must be an attracting spiral point. According to a theorem of Coppel [4], there is no periodic orbit for this quadratic dynamical system because the separatrix leaves the saddle-node at infinity. Thus the orbit through each point in the phase plane must spiral toward C.

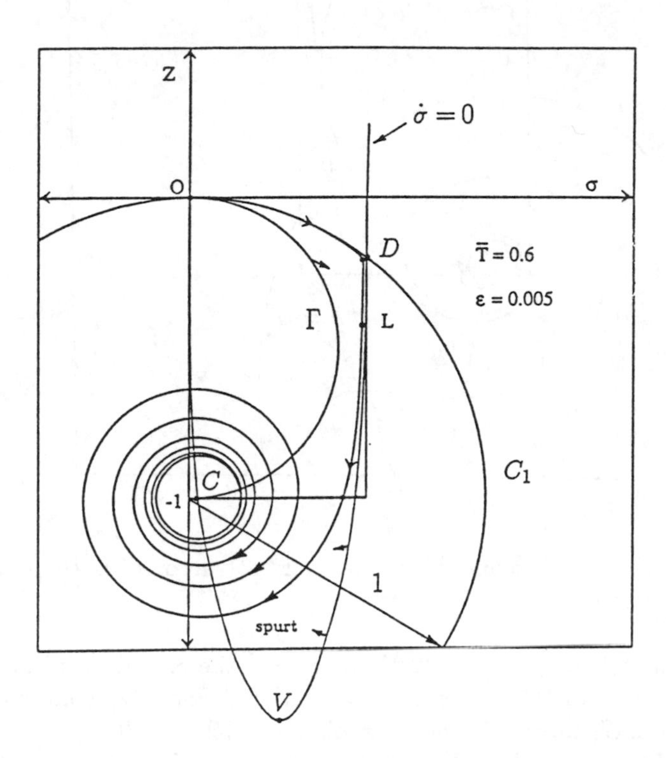

Fig. 6: The orbit through origin when the spurt attractor C is the only critical point.

In the application to the shear flow problem, we are interested in the particular solution of Eq. (3.1) with initial data $\sigma = 0$ and $Z = 0$ (point O). This solution initially remains inside the region $\mathcal{R}$ bounded by $ODVCO$ in Fig. 6, eventually exits through the arc CV of the parabola, and finally spirals toward C. Indeed, the orbit through O must remains inside C_1 and outside Γ because of Eq. (3.10); and for points along the portion of the parabola between D and V, $\dot{Z} = 0$ and $\dot{\sigma} < 0$, so that orbits cannot leave through DV. Therefore it must leave $\mathcal{R}$ along the arc CV of the parabola, whereupon it spirals into C. This solution is illustrated in Fig. 6.

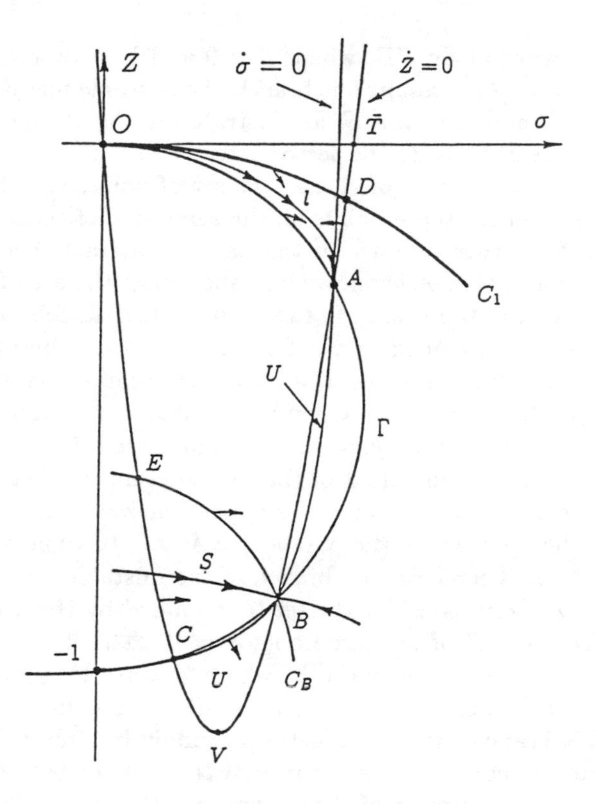

Fig. 7: Invariant regions in the case of three critical points.

B.

Next consider the case when there are three critical points, illustrated in Fig. 7. As shown by A. Coppel [private communication] using the Bendixson criterion, there are no periodic orbits or separatrix cycles for this quadratic dynamical system. Therefore, as t approaches infinity, the orbit through any point in the plane either: tends to A; tends to C; or tends to B along its stable manifold.

We first prove that the closed set $\mathcal{I}$ bounded by the curved triangle OAD is invariant with respect to Eqs. (3.1). To this end we show that orbits starting from points along OAD remain in $\mathcal{I}$. For points along C_1 strictly between O and D, this follows from the invariance of C_1. For points along the portion of the parabola between A and D (excepting A), $\dot{Z} = 0$ and $\dot{\sigma} < 0$, so that orbits lead into $\mathcal{I}$. Similarly, the flow leads into $\mathcal{I}$ along the arc of Γ strictly between A and O; this is because $\dot{Z} < 0$ and because of Eq. (3.10). Finally, A is a critical point, while the orbit through O must remains inside C_1 and outside Γ because of Eq. (3.10). One consequence of the invariance of $\mathcal{I}$ is that the solution of Eq. (3.1) with initial data $\sigma = 0$ and $Z = 0$ flows into the classical attractor A.

Next we study the stable manifold for the saddle point B in Fig. 7. Through this point we have drawn the circle C_B centered at $\sigma = 0$ and $Z = -1$, which intersects the parabola at E. Let $\mathcal{S}$ denote the closed set bounded by the curved triangle ECB. At points on the boundary, the flow is directed as follows: outward from $\mathcal{S}$ along EB

because of Eq. (3.10); outward along CB, where $\dot{Z} < 0$ and $\dot{V} = 0$; and inward along CE because $\dot{\sigma} > 0$ and $\dot{Z} = 0$. As a result, one branch of the stable manifold at B must enter $\mathcal{S}$ through the arc EC and remain is $\mathcal{S}$, as illustrated in Fig. 7. The other branch of the stable manifold enters B through the sector exterior to Γ and the circle C_B.

Notice that the basin of attraction of A, *i.e.*, the set of points that flow toward A as t approaches infinity, comprises those points on the same side of the stable manifold of B as is A; points on the other side are in the basin of attraction of C. For the purpose of analyzing the spurt phenomenon, we now show that the arc of Γ between B and the origin O is contained in the basin of attraction of A. This follows because the flow is directed into the region bounded by the following curves: Γ between B and A; the parabola between A and D; C_1 between D and O; the parabola between O and E; and C_B between E and B. Therefore this region is invariant. In particular, the stable manifold for B cannot cross the boundary, so that it cannot cross Γ between B and O.

Finally, consider the unstable manifold of the saddle point B. Let $\mathcal{U}_1$ be the set bounded by the arcs of the parabola $\dot{Z} = 0$ and the hyperbola $\dot{\sigma} = 0$ between the critical points B and A. Along the open arc of the parabola BA, $\dot{\sigma} > 0$, while along the open arc of the hyperbola, $\dot{Z} > 0$. Therefore, one branch of the unstable manifold at B lies in $\mathcal{U}_1$ and connects B to A. Next consider the set $\mathcal{U}_2$ bounded by the arc BVC of the parabola $\dot{Z} = 0$ and the arc CB of the hyperbola $\dot{\sigma} = 0$. The flow is directed into $\mathcal{U}_2$ both along BV, where $\dot{\sigma} < 0$, and along CB, where $\dot{Z} < 0$. Therefore, the second branch of the unstable manifold at B remains in $\mathcal{U}_2$ until it exits through the arc VC. If C is a spiral point, this branch enters and leaves $\mathcal{U}_2$ infinitely often as it spirals into C, while if C is an attracting node, it does not reenter $\mathcal{U}_2$ as it tends to C.

To summarize the above description of the dynamics of the system (3.1) in the case of three critical points, with C being a spiral point, the reader is referred to Fig. 8.

C.

Finally consider the case of a single critical point at A, which is an attracting node. For quadratic dynamical systems, a periodic orbit must enclose a center or a spiral point [3]; thus there is no periodic orbit. As a result, all orbits are attracted to the node at A. The orbit through the origin remains in a region that is analogous to the region $\mathcal{I}$ in Fig. 7.

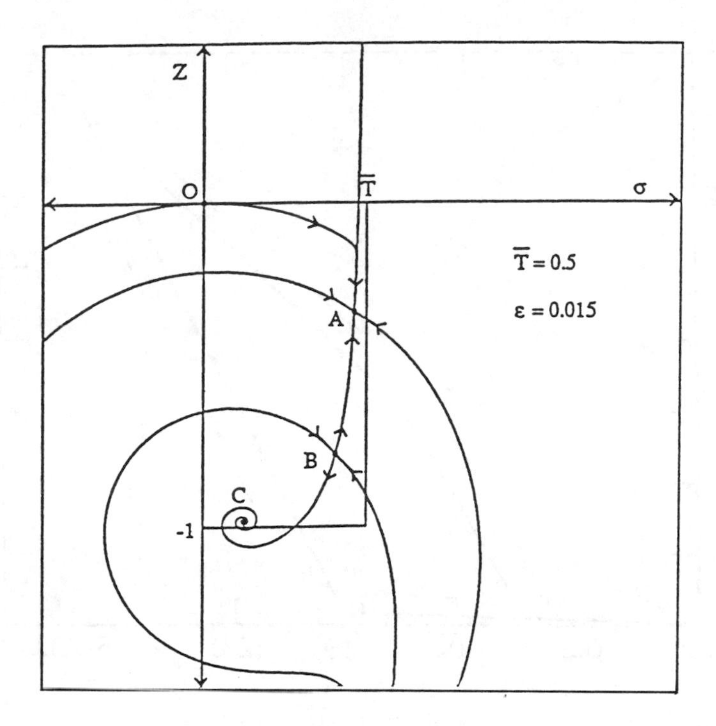

Fig. 8: Phase portrait in the case of three critical points, with C being a spiral.

4. Features of the Mathematical Model in Relation to Experiment

The numerical simulations of (JS) described in Refs. [9, 13] exhibited several effects related to spurt: latency, shape memory, and hysteresis. For example, Fig. 9 shows the result of simulating a loading sequence in which the pressure gradient $\overline{f}$ is increased in small steps, allowing sufficient time between steps to achieve steady flow [9]. The loading sequence is followed by a similar unloading sequence, in which the driving pressure gradient is decreased in steps. The initial step used zero initial data, and succeeding steps used the results of the previous step as initial data. The resulting hysteresis loop includes the shape memory described in Ref. [7] for a simpler model. The width of the hysteresis loop at the bottom can be related directly to the molecular weight of the sample [9].

In this section we explain these effects using the results of the phase plane analysis of the dynamical system (3.1). We consider experiments of the following type: the flow is initially in a steady state corresponding to a forcing $\overline{f}$, and the forcing is suddenly changed to $\overline{f} + \Delta\overline{f}$. We call this process "loading" (resp. "unloading") if $\Delta\overline{f}$ has the same (resp. opposite) sign as $\overline{f}$.

Let us first establish some convenient terminology. Given a value of $\overline{f}$, the channel $-1/2 \leq x \leq 0$ can be subdivided into three contiguous zones (subintervals) according to the size of $\overline{T}(x) = -\overline{f}x$. (Refer to Figs. 2 and 3.) We define Zone 1, which is nearest

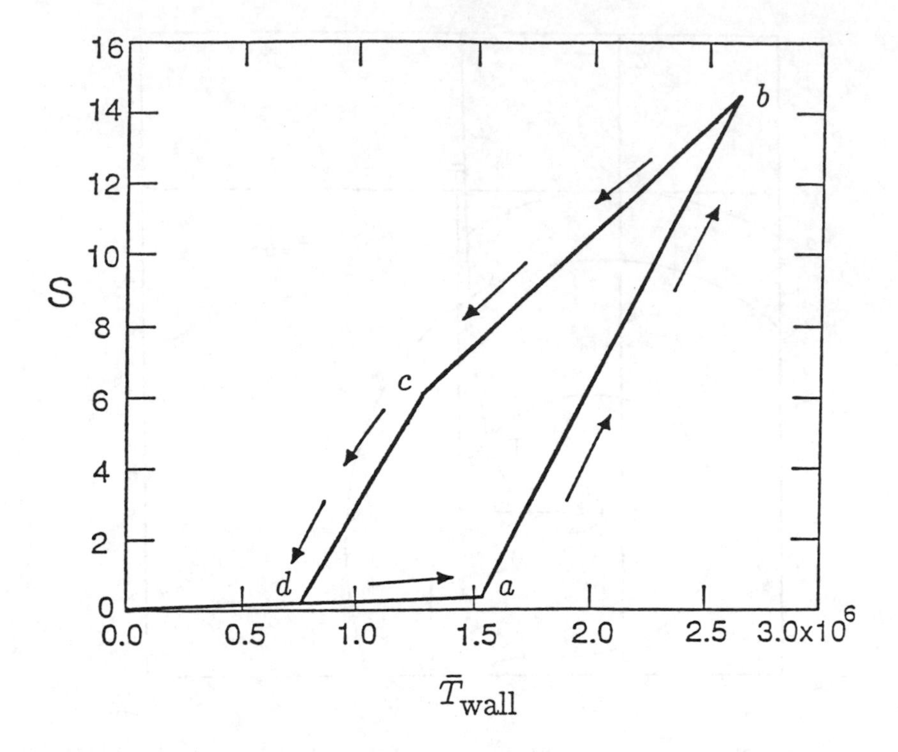

Fig. 9: Hysteresis under cyclic load: normalized throughput S *vs.* wall shear stress $\overline{T}_{\text{wall}}$ [9].

to the wall $x = -1/2$, to comprise those points x for which $\overline{T}(x) \geq \overline{T}_M$; this subinterval is nonempty if $\overline{f}$ is supercritical, *i.e.*, $\overline{f}/2 > \overline{T}_M$. For points in Zone 1, the only critical point of the system (3.1) is C, as in Fig. 5. In Zone 2, where $\overline{T}_m \leq \overline{T}(x) \leq \overline{T}_M$, there are three critical points, A, B, and C, as in Fig. 7; this subinterval is nonempty if $\overline{f}/2 > \overline{T}_m$. Zone 3, which is nearest to the centerline $x = 0$, consists of x for which $\overline{T}(x) \leq \overline{T}_m$; the corresponding phase plane has only A as a critical point.

Notice also that the critical points for Eq. (3.1), with any value of $\overline{T}$, lie on the circle Γ, which is independent of $\overline{T}$. Let (σ_M, Z_M) denote the degenerate (double root) critical point that occurs when $\overline{T} = \overline{T}_M$, *i.e.*, at "top jumping" in Fig. 2; and let (σ_m, Z_m) denote the degenerate critical point for $\overline{T} = \overline{T}_m$, *i.e.*, for "bottom jumping." These points serve to divide Γ into arcs: Γ_A, the upper arc of Γ between (σ_M, Z_M) and $(-\sigma_M, Z_M)$; Γ_C, the lower arc between (σ_m, Z_m) and $(-\sigma_m, Z_m)$; and Γ_B, the remaining two arcs where $Z \in [Z_m, Z_M]$. For any value of $\overline{T}$, positive or negative, the classical attractor A lies in Γ_A, the spurt attractor C lies in Γ_C, and the saddle point B lies in Γ_B. Furthermore, as $|\overline{T}|$ is increased, the critical points A and C move downward along Γ, while B moves upward. This follows from Eq. (3.3) by differentiating the relation $Q(\sigma/\overline{T}) = 0$, to determine how $\sigma/\overline{T}$ varies with $\overline{T}$, and by using the first of Eqs. (3.2).

A. Startup

As a first experiment, consider starting from the quiescent state at the origin $\sigma = 0$, $Z = 0$ and loading to $\overline{f} > 0$. For each x in Zones 2 and 3 (near the centerline), the origin in the corresponding phase plane lies in the basin of attraction of the node at A, so that the orbit through the origin tends to A; this is illustrated in Fig. 8. For each x in Zone 1, by contrast, the origin is attracted to the sole critical point at C, as in Fig. 6. Accordingly, we draw two conclusions: (a) if $\overline{f}$ is subcritical, the flow approaches the classical solution corresponding to A at every point x; (b) if $\overline{f}$ is supercritical, the flow approaches a steady spurt solution in which the jump in strain rate occurs at the shear stress maximum $\overline{T}_M$ (which is top jumping in Fig. 2), *i.e.*, such that the kink in the velocity profile (see Fig. 3) is located as close as possible to the wall.

B. Loading

Next, consider increasing the load from a supercritical value $\overline{f} > 0$ to $\overline{f}_1 > \overline{f}$. This causes the three zones to shift: some points x previously in Zone 2 or 3 for $\overline{f}$ now lie in Zone 1 for $\overline{f}_1$. For each x in the new Zone 1, the corresponding phase plane for the system (3.1) has a unique critical point C_1, the spiral point. If such an x was previously in Zone 1 or 2, C_1 lies further down along Γ_C than was the corresponding attractor C for the smaller load $\overline{f}$. Similarly, for each x in the new Zone 2, there are three critical points, A_1, B_1, and C_1. Again, the spurt attractor C_1 lies further down along Γ_C, while the saddle point B_1 lies further up along Γ_B and the classical attractor A_1 lies further down along Γ_A, as compared to the corresponding critical points for $\overline{f}$. In particular, the stable manifold of B_1 lies above that of B, at least near to B_1. Finally, in the new Zone 3, there is a single critical point A_1 located downward along Γ_A with respect to A.

Let us now determine how the steady profile changes as a result of the increased load. For every $x \in [-\frac{1}{2}, 0]$, we take the steady solution (σ_0, Z_0) attained at the load $\overline{f}$ as the initial point for an orbit in the phase portrait for the new load $\overline{f}_1$. If x belongs to the new Zone 1, the initial point (σ_0, Z_0) is either a classical attractor A, which disappeared as the loading was increased, or a spurt attractor C. Regardless, the orbit through this point leads to C_1 as time progresses because C_1 is the only critical point for the load $\overline{f}_1$. Thus spurt continues throughout the new Zone 1. By contrast, if x lies the new Zone 2 or 3, the initial point (σ_0, Z_0), which is a classical attractor A on Γ_A between A_1 and the origin, lies in the basin of attraction of the node A_1, and the corresponding orbit tends to A_1. To see this for points in Zone 2, notice that neither branch of the stable manifold of B_1 can intersect Γ between B_1 and the origin, as was shown in Sec. 3. Thus the domain of attraction of A_1, which is bounded by the stable manifold of B_1, contains A.

As a result, a point in x in the channel can change only from a classical attractor to a spurt attractor, and then only if $\overline{T}_1(x) = -\overline{f}_1 x$ exceeds $\overline{T}_M$. In other words, loading causes the position x_* of the kink in Fig. 3 to move away from the wall, but only to the extent that it must. (Formulas for the precise location are found in Refs. [9, 14].) Therefore, a loading process, without an intervening unloading, yields a jump in strain rate at total stress $\overline{T}_M$, *i.e.*, top jumping.

C. Latency

Our next task is to explain the latency effect that occurs during loading. In this context we assume that ε is small. It follows from Eqs. (2.5) and (2.6) that the total stress $\overline{T}_M$ at the the local maximum M is $1/2 + O(\varepsilon)$, while the local minimum m corresponds to a total stress $\overline{T}_m$ of $2\sqrt{\epsilon}\,[1+O(\epsilon)]$. Furthermore, for x such that $\overline{T}(x) = O(1)$, $\sigma = \overline{T} + O(\varepsilon)$ at an attracting node at A, while $\sigma = O(\epsilon)$ at a spurt attractor C (which is a spiral). Consider a point along the channel for which $\overline{T}(x) > \overline{T}_M$, so that the only critical point of the system (3.1) is C, and suppose that that $\overline{T} < 1$. Then the evolution of the system exhibits three distinct phases, as indicated in Fig. 6: an initial "Newtonian" phase (O to N); an intermediate "latency" phase (N to S); and a final "spurt" phase (S to C).

The Newtonian phase occurs on a time scale of order ε, during which the system approximately follows an arc of a circle centered at $\sigma = 0$ and $Z = -1$. Having assumed that $\overline{T} < 1$, Z approaches

$$Z_N = (1 - \overline{T}^2)^{\frac{1}{2}} - 1 \tag{4.1}$$

as σ rises to the value $\overline{T}$. (If, on the other hand, $\overline{T} \geq 1$, the circular arc does not extend as far as $\overline{T}$, and σ never attains the value $\overline{T}$; rather, the system slowly spirals toward the spurt attractor. Thus the dynamical behavior does not exhibit distinct phases.)

The latency phase is characterized by having $\sigma = \overline{T} + O(\epsilon)$, so that σ is nearly constant and Z evolves approximately according to the differential equation

$$\dot{Z} = -\frac{\overline{T}^2}{Z+1} - Z\,. \tag{4.2}$$

Therefore, the shear stress and velocity profiles closely resemble those for a steady solution with no spurt, but the solution is not truly steady because the normal stress difference Z still changes. Integrating Eq. (4.2) from $Z = Z_N$ to $Z = -1$ determines the latency period. This period becomes indefinitely long when the forcing decreases to its critical value; thus the persistence of the near-steady solution with no spurt can be very dramatic. The solution remains longest near point L where $Z = -1 + \overline{T}$. This point may be regarded as the remnant of the attracting node A and the saddle point B.

Eventually the solution enters the spurt phase and tends to the critical point C. Because C is an attracting spiral, the stress oscillates between the shear and normal components while it approaches the steady state.

D. Unloading: Shape Memory and Hysteresis

Now consider unloading from a steady solution for the load $\overline{f}_1$ to the load $\overline{f} < \overline{f}_1$; assume, for the moment, that $\overline{f}$ and $\overline{f}_1$ are both positive. The initial steady solution need not correspond to top jumping, as would be obtained by a pure loading process. Again the zones shift: some points previously in Zone 1 move into the new Zone 2 or 3. The orbit through any point in the new Zone 1 tends to the spurt attractor C for $\overline{f}$, and the orbit through any point in the new Zone 3 tends to the classical attractor A. More generally, the orbit through any point starting at a classical attractor A_1 leads to a new classical attractor A; this follows, as before, because the stable manifold for B cannot cross Γ between B and the origin. For points in the new Zone 2 that initiate at a spurt attractor C_1, however, there is an apparent choice of final rest state.

Clearly, the answer depends on whether C_1 lies on the same side of the stable manifold through B as does C. If this is true, the orbit through this point tends to C, so that spurt continues at this point. Suppose, for example, that it is true of all points in the new Zone 2 that initiate at a spurt attractor. Then all points that were classical remain classical, and spurt continues at all other points. Thus the position x_* of any jump in strain rate stays fixed, even though other flow characteristics (such as the magnitudes of the velocities) change. This phenomenon was termed "shape memory" by Hunter and Slemrod [7].

As an instance of this, suppose that the stress $\overline{T}(x)$ of smallest magnitude for points x in the spurt layer is strictly greater than $\overline{T}_m$. Then if $\overline{f}$ is sufficiently close to $\overline{f}_1$, $\overline{T}(x)$ for each x in the layer remains larger than $\overline{T}_m$; thus no point in the layer belongs to the new Zone 3. Moreover, for such an x, the stable manifold through B is only a slight perturbation of the stable manifold through B_1, which surrounds C_1, so that it surrounds C. Thus shape memory occurs in the situation where the stress in the layer is separated by a gap from $\overline{T}_m$ and the loading increment is sufficiently small. Because the stress at a jump in strain rate falls in the open interval $\overline{T}_m < \overline{T} < \overline{T}_M$, such solutions are referred to as "intermediate jumping;" this is the case illustrated in Fig. 2.

If, on the other hand, the minimum stress in the layer is $\overline{T}_m$ so that one layer boundary corresponds to bottom jumping, then the spurt layer must shift upon unloading. Indeed, there are points x in the layer for which $\overline{T}(x) < \overline{T}_m$, so that they have moved into the new Zone 3 and must flow to a classical attractor. The layer moves to the point x_* farthest from the wall such that $\overline{T}(x_*) = \overline{T}_m$. Similarly, shape memory is lost if $\overline{f}$ is lowered enough that $\overline{T}(x)$ drops below $\overline{T}_m$ in the layer. It is also possible, when the change in loading is large, for spurt attractors C_1 to lie on the opposite side of the stable manifold of B from C. This causes the formation of a region of classical flow next to the wall, which can coexist with intermediate spurt layers as well as the classical flow in the center of the channel.

This picture of shape memory explains the hysteresis loop in Fig. 9 obtained in a loading–unloading sequence. In this figure, the throughput S, which is proportional to the area under the velocity profile in Fig. 3, is plotted as a function of the wall shear stress $\overline{T}_{\text{wall}} = \overline{f}/2$. The portion of this curve between the origin and a corresponds to subcritical loading, $\overline{T}_{\text{wall}} < \overline{T}_M$, while the segment ab corresponds to top jumping in supercritical loading. Unloading commences at b and continues along bc and cd. The throughput along bcd is different from the loading curve because intermediate and bottom jumping solutions occur along the unloading curve. In fact, the layer does not move during the process bc because of shape memory, so that the spurt layer is wider during unloading, resulting in larger throughput. At some point between b and c, the imposed $\overline{T}_{\text{wall}}$ becomes lower than $\overline{T}_M$; yet the flow remains supercritical. In such a situation, the flow would be subcritical at the corresponding stress in loading, and the difference in throughput in loading and unloading is particularly dramatic. At point c, bottom jumping commences, and the throughput decreases much more rapidly because the layer moves toward the wall. This accounts for the discontinuity in slope at c.

The salient features of this explanation of hysteresis are: the hysteresis loop opens from the point at which unloading commences; no part of the unloading path retraces the loading path until point d; and there is a discontinuity in slope of the unloading portion of the loop. These features stand in marked contrast to other plausible predictions of the nature of the hysteresis in spurt [15]; experiments are needed to verify which theory is correct.

E. Reloading and Flow Reversal

It is possible, of course, to apply more complex load sequences than those just described. For example, a loading sequence can be followed by unloading, which, in turn, is followed by reloading. Arguments extending the ones given above can be used to make qualitative predictions in such cases. For example, the layer position can remain fixed upon reloading.

Another striking phenomenon predicted by this analysis occurs when $\overline{f}_1$ and $\overline{f}$ have opposite signs and $|\overline{f}| < |\overline{f}_1|$. This more general form of unloading (defined in Ref. [9]) causes a reversal of the flow direction; surprisingly, though, the layer position can remain unchanged. For this to happen, the magnitude of the stress at the layer boundary must not fall below $\overline{T}_m$, and the stable manifolds of all saddle points B (which are now in the left half-plane of the phase portrait) must extend far enough into the right half-plane to enclose the spurt attractors C_1. For small ε, the stable manifold for a saddle point B is nearly a circle, and shape memory in flow reversal is more the rule than the exception. Because we doubted our observation of shape memory during flow reversals in numerical simulations, we were motivated to pursue the the more rigorous analysis described above.

F. Summary

To summarize, the phenomena of top jumping upon loading and shape memory and hysteresis upon unloading follow from analyzing the phase portraits of the approximating system (3.1). The analysis reduces to asking, for each point in the channel, whether or not the steady state for the initial load lies in the basin of attraction of the classical attractor for the new load. More complicated load sequences can be analyzed easily by answering this question.

5. Conclusions

The phase plane analysis of the approximating dynamical system (3.1) accurately reproduces the spurt behavior in viscoelastic shear flows that was observed experimentally by Vinogradov *et al.* [20] and numerically in Refs. [9, 12, 13]. Furthermore, this analysis predicts several associated phenomena, also observed numerically, such as latency, hysteresis, and shape memory; rheological experiments to verify these effects would be valuable.

Acknowledgments

We thank Professor A. Coppel for an elegant argument ruling out the existence of periodic and separatrix cycles for the system (3.1). We also acknowledge helpful discussions with D. Aronson, G. Sell, M. Slemrod and A. Tzavaras, and we thank M. Yao for help with the figures.

References

1. R. Bird, R. Armstrong, and O. Hassager, *Dynamics of Polymeric Liquids*, John Wiley and Sons, New York, 1987.

2. E. Coddington and N. Levinson, *Theory of Ordinary Differential Equations*, McGraw-Hill, New York, 1955.

3. A. Coppel, "A Survey of Quadratic Systems," *J. Differential Equations* **2** (1966), pp. 293–304.

4. A. Coppel, "A Simple Class of Quadratic Systems," *J. Differential Equations* **64** (1986), pp. 275–282.

5. M. Doi and S. Edwards, "Dynamics of Concentrated Polymer Systems," *J. Chem. Soc. Faraday* **74** (1978), pp. 1789–1832.

6. C. Guillopé and J.-C. Saut, "Global Existence and One-Dimensional Nonlinear Stability of ShearingMotions of Viscoelastic Fluids of Oldroyd Type," *Math. Mod. Numer. Anal.*, 1989. To appear.

7. J. Hunter and M. Slemrod, "Viscoelastic Fluid Flow Exhibiting Hysteretic Phase Changes," *Phys. Fluids* **26** (1983), pp. 2345–2351.

8. M. Johnson and D. Segalman, "A Model for Viscoelastic Fluid Behavior which Allows Non-Affine Deformation," *J. Non-Newtonian Fluid Mech.* **2** (1977), pp. 255–270.

9. R. Kolkka, D. Malkus, M. Hansen, G. Ierley, and R. Worthing, "Spurt Phenomena of the Johnson-Segalman Fluid and Related Models," *J. Non-Newtonian Fluid Mech.* **29** (1988), pp. 303–325.

10. R. Kolkka and G. Ierley, "Spurt Phenomena for the Giesekus Viscoelastic Liquid Model," F.R.O.G.-TR 88-20, Michigan Technical Univ., Houghton, MI, 1989.

11. Y.-H. Lin, "Explanation for Slip-Stick Melt Fracture in Terms of Molecular Dynamics in Polymer Melts," *J. Rheol.* **29** (1985), pp. 609–637.

12. D. Malkus, J. Nohel, and B. Plohr, "Time-Dependent Shear Flow of a Non-Newtonian Fluid," in *Current Progress in Hyperbolic Systems: Riemann Problems and Computations (Bowdoin, 1988)*, Contemporary Mathematics, ed. B.-Lindquist, American Mathematics Society, Providence, RI, 1989. To appear.

13. D. Malkus, J. Nohel, and B. Plohr, "Dynamics of Shear Flow of a Non-Newtonian Fluid," *J. Comput. Phys.*, 1989. To appear.

14. D. Malkus, J. Nohel, and B. Plohr, "Phase-Plane and Asymptotic Analysis of Spurt Phenomena," in preparation, 1989.

15. T. McLeish and R. Ball, "A Molecular Approach to the Spurt Effect in Polymer Melt Flow," *J. Polymer Sci.* **24** (1986), pp. 1735–1745.

16. J. Nohel, R. Pego, and A. Tzavaras, "Stability of Discontinuous Shearing Motions of Non-Newtonian Fluids," in preparation, 1989.

17. J. Oldroyd, "Non-Newtonian Effects in Steady Motion of Some Idealized Elastico-Viscous Liquids," *Proc. Roy. Soc. London* **A 245** (1958), pp. 278–297.

18. B. Plohr, "Instabilities in Shear Flow of Viscoelastic Fluids with Fading Memory," in *Workshop on Partial Differential Equations and Continuum Models of Phase Transitions (Nice, 1988)*, ed. D. Serre, Springer–Verlag, New York, 1988. Lecture Notes in Mathematics, to appear.

19. M. Renardy, W. Hrusa, and J. Nohel, *Mathematical Problems in Viscoelasticity*, Pitman Monographs and Surveys in Pure and Applied Mathematics, Vol. 35, Longman Scientific & Technical, Essex, England, 1987.

20. G. Vinogradov, A. Malkin, Yu. Yanovskii, E. Borisenkova, B. Yarlykov, and G. Berezhnaya, "Viscoelastic Properties and Flow of Narrow Distribution Polybutadienes and Polyisoprenes," *J. Polymer Sci., Part A-2* **10** (1972), pp. 1061–1084.

For additional references see the Appendix

BOUNDARY CONDITIONS FOR STEADY FLOWS OF VISCOELASTIC FLUIDS

Michael Renardy
Department of Mathematics and ICAM, Virginia Tech
Blacksburg, VA 24061-0123, USA

Abstract

We discuss small perturbations of uniform flow of a viscoelastic fluid transverse to a strip. The constitutive relation is assumed to be of Maxwell or Jeffreys type. We present a summary of recent results concerning the choice of boundary conditions which lead to a well-posed problem.

1. Introduction

While the study of existence and uniqueness results for steady flows of Newtonian fluids is well advanced, relatively little is known about viscoelastic fluids with memory. For such fluids, the nature of boundary conditions leading to well-posed problems is in general different from the Newtonian case. There are two reasons for this:

1. The memory of the fluid implies that what happens in the domain under consideration is dependent on the deformation history of the fluid before it entered the domain. Information about this deformation history must therefore be given in the form of boundary conditions at inflow boundaries. The precise nature of such inflow conditions is dependent on the constitutive relation; in general, an infinite numbers of conditions would be needed. In the following, we shall only consider differential models of Maxwell or Jeffreys type. Such models are frequently used in numerical simulations.
2. For fluids of Maxwell type, there is a change of type in the governing equations when the velocity of the fluid exceeds the propagation speed of shear waves (cf. [2], [3], [10], [11]). This necessitates a change in the nature of boundary conditions. If boundary conditions which would be correct in the subcritical case are imposed in a supercritical situation, the problem becomes ill-posed in a similar fashion as the Dirichlet problem for the wave equation (see [7]).

The equations considered in the following are the balance of momentum,

$$\rho(v \cdot \nabla)v = \mu\Delta v + \text{div } \mathbf{T} - \nabla p + f, \tag{1}$$

the incompressibility condition,

$$\text{div } v = 0, \tag{2}$$

and a differential equation relating $\mathbf{T}$ to the motion,

$$(v \cdot \nabla)\mathbf{T} + \lambda\mathbf{T} + \mathbf{g}(\nabla v, \mathbf{T}) = \eta\lambda(\nabla v + (\nabla v)^T). \tag{3}$$

The unknowns are the velocity v, the pressure p and the viscoelastic stress tensor $\mathbf{T}$. The density ρ and the quantities η and λ are given positive constants. The constant μ is a Newtonian contribution to the viscosity, which is either positive or zero. If μ is positive, we call the fluid "of Jeffreys type", if it is zero, the fluid is "of Maxwell type". The body force f is prescribed, and $\mathbf{g}$ is a given nonlinear function such that g and its first derivatives vanish when the arguments are zero. The exact form of $\mathbf{g}$ depends on the specific model. A number of models of Maxwell and Jeffreys type were introduced by Oldroyd [4], and many others have been proposed since.

We seek solutions of (1)-(3) in the strip $0 \leq x \leq 1$, with periodic boundary conditions imposed in the y- and z-directions; the periods are denoted by L and M. The solutions we seek are small perturbations of uniform flow given by $v = (V, 0, 0)$, $p = 0$, $\mathbf{T} = 0$, where $V > 0$. The given body force f and any prescribed boundary conditions are assumed to satisfy smallness conditions consistent with this. The analysis employs function spaces of Sobolev type; we denote spaces of functions defined on the strip by H^s, and spaces of functions defined on one of the boundaries by $H^{\langle s \rangle}$; the corresponding norms are denoted by $\|\cdot\|_s$ and $\|\cdot\|_{\langle s \rangle}$, respectively.

2. Fluids of Jeffreys type

For fluids of Jeffreys type, it is possible to prescribe the velocities on the boundary, as well as the values of $\mathbf{T}$ at the inflow boundary. If the inertial terms in the momentum equation are neglected, we can also give a theorem for the case when tractions rather than velocities are prescribed. The analysis can be based on energy estimates in conjunctions with a parabolic regularization of (3) (i.e. a term $\epsilon(\frac{\partial^2}{\partial y^2} + \frac{\partial^2}{\partial z^2})\mathbf{T}$ is added to the right hand side of (3)). Details for the case of traction conditions can be found in [8]; the same method also works for velocity conditions. An alternative proof for velocity conditions is sketched in [6]. In the following, we shall only state the results.

Let us first consider the case of velocity boundary conditions. We prescribe

$$v(0, y, z) = (V, 0, 0) + v_0(y, z),\ v(1, y, z) = v_1(y, z),\ \mathbf{T}(0, y, z) = \mathbf{T}_0(y, z), \tag{4}$$

where

$$\int_0^M \int_0^L e_x \cdot v_0(y, z)\, dy\, dz = \int_0^M \int_0^L e_x \cdot v_1(y, z)\, dy\, dz. \tag{5}$$

The following theorem holds.

Theorem 1:

Assume that $\|f\|_1$, $\|v_0\|_{\langle 5/2 \rangle}$, $\|v_1\|_{\langle 5/2 \rangle}$ and $\|\mathbf{T}_0\|_{\langle 2 \rangle}$ are sufficiently small. Then there exists a solution of (1)-(3) subject to the given boundary conditions such that $v \in H^3$, $p \in H^2$, $\mathbf{T} \in H^2$. Except for a constant in the pressure, this solution is the only one for which $\|v - (V, 0, 0)\|_3 + \|\mathbf{T}\|_2$ is small.

For the case of traction conditions, we assume that the inertial terms in (1) are neglected, i.e. we set $\rho = 0$. At the inflow boundary $x = 0$, we prescribe the components of $\mathbf{T}$,

$$\mathbf{T} = \mathbf{T}_0, \tag{6}$$

as well as the tractions

$$T_{11} + 2\mu\frac{\partial v_1}{\partial x} - p = t_1,\ T_{12} + \mu(\frac{\partial v_2}{\partial x} + \frac{\partial v_1}{\partial y}) = t_2,\ T_{13} + \mu(\frac{\partial v_3}{\partial x} + \frac{\partial v_1}{\partial z}) = t_3. \tag{7}$$

At the outflow boundary $x = 1$ we prescribe the tractions:

$$T_{11} + 2\mu\frac{\partial v_1}{\partial x} - p = s_1,\ T_{12} + \mu(\frac{\partial v_2}{\partial x} + \frac{\partial v_1}{\partial y}) = s_2,\ T_{13} + \mu(\frac{\partial v_3}{\partial x} + \frac{\partial v_1}{\partial z}) = s_3. \tag{8}$$

The body force and tractions have to satisfy the consistency condition

$$\int_0^L \int_0^M s - t\ dz\ dy + \int_0^1 \int_0^L \int_0^M f\ dz\ dy\ dx = 0. \tag{9}$$

Since these boundary conditions can determine the velocity only up to a constant, we prescribe the average velocity:

$$\int_0^1 \int_0^L \int_0^M v\ dz\ dy\ dx = LM(V + \alpha, \beta, \gamma), \tag{10}$$

where α, β, and γ are given numbers. The following result holds.

Theorem 2:

Assume that $\|f\|_1$, $\|\mathbf{T}_0\|_{\langle 2\rangle}$, $\|s\|_{\langle 3/2\rangle}$, $\|t\|_{\langle 3/2\rangle}$ and $|\alpha| + |\beta| + |\gamma|$ are sufficiently small. Then there exists a solution of (1)-(3) subject to the given boundary conditions such that $v \in H^3$, $p \in H^2$ and $\mathbf{T} \in H^2$. This solution is the only one for which $\|v - (V, 0, 0)\|_3 + \|\mathbf{T}\|_2$ is small.

3. Maxwell fluid, subcritical case

We now assume that $\mu = 0$. We apply the operation $(v \cdot \nabla) + \lambda + (\nabla v)^T$ to equation (1), and we use (3) to reexpress $((v \cdot \nabla) + \lambda)\mathbf{T}$. This results in an equation of the form

$$\rho V^2 \frac{\partial^2 v}{\partial x^2} + \rho\lambda\frac{\partial v}{\partial x} = \eta\lambda\Delta v - \nabla q + h(v, \nabla v, \nabla^2 v, \mathbf{T}, \nabla\mathbf{T}). \tag{11}$$

Here h contains terms of quadratic and higher order, and we have set $q = (v \cdot \nabla)p + \lambda p$. If $\rho V^2 < \eta\lambda$ (we refer to this as the subcritical case), then the system consisting of (11) and (2) is elliptic and results analogous to those for the Stokes equation apply. The construction of solutions is now based on iterating between (11), (2) and (3). That is, we use the iteration

$$\rho V^2 \frac{\partial^2 v^{n+1}}{\partial x^2} + \rho\lambda\frac{\partial v^{n+1}}{\partial x} = \eta\lambda\Delta v^{n+1} - \nabla q^{n+1} + h(v^n, \nabla v^n, \nabla^2 v^n, \mathbf{T}^n, \nabla\mathbf{T}^n),$$

$$\text{div } v^{n+1} = 0, \tag{12}$$

$$(v^{n+1} \cdot \nabla)\mathbf{T}^{n+1} + \lambda \mathbf{T}^{n+1} + \mathbf{g}(\nabla v^{n+1}, \mathbf{T}^n) = \eta\lambda(\nabla v^{n+1} + (\nabla v^{n+1})^T). \tag{13}$$

Iterations of this nature have been use successfully in numerical simulations (see e.g. [1]). An appropriate boundary condition for (12) is the prescription of velocities on the boundary, and an appropriate condition for (13) is the prescription of $\mathbf{T}$ at the inflow boundary. However, it would be wrong to prescribe those conditions. The reason is that we have substituted (11) for (1), and while (1) implies (11), the converse is not true. To insure that (1) actually holds, we must impose (1) at the inflow boundary. That is, we require that at $x = 0$, we must have

$$\rho(v^{n+1} \cdot \nabla)v^{n+1} = \text{div } \mathbf{T}^{n+1} - \nabla p^{n+1} + f. \tag{14}$$

This, in conjunction with (13) and the equation

$$(v^{n+1} \cdot \nabla)p^{n+1} + \lambda p^{n+1} = q^{n+1} \tag{15}$$

can be used to express some components of $\mathbf{T}^{n+1}$ in terms of others. Not all components of $\mathbf{T}^{n+1}$ can therefore be prescribed. It turns out that in two space dimensions, it is possible to prescribe the diagonal components of $\mathbf{T}$, while T_{12} can be determined at each step of the iteration from (13)-(15). In three dimensions, the situation is more complicated. At the inflow boundary, we expand each stress component in a Fourier series, e.g.

$$T_{11}(0, y, z) = \sum_{k,l} t_{11}^{kl} \exp(2\pi i(ky/L + lz/M)). \tag{16}$$

Then one can, for example, prescribe the following inflow conditions:

$$\begin{aligned} &t_{11}^{kl},\ t_{22}^{kl},\ t_{13}^{kl},\ t_{33}^{kl} \text{ if } |k| >> |l|, \\ &t_{11}^{kl},\ t_{22}^{kl},\ t_{12}^{kl},\ t_{33}^{kl} \text{ if } |l| >> |k|, \\ &t_{11}^{kl},\ t_{13}^{kl},\ t_{23}^{kl},\ t_{33}^{kl} \text{ if } |k| \text{ and } |l| \text{ are comparable}, \\ &t_{11}^{kl},\ t_{22}^{kl},\ t_{23}^{kl},\ t_{33}^{kl} \text{ if } k = l = 0. \end{aligned} \tag{17}$$

Let $\mathbf{T}_p$ denote the prescribed part of the stress according to (17) and, in addition, let us prescribe the velocity

$$v(0, y, z) = (V, 0, 0) + v_0(y, z),\ v(1, y, z) = (V, 0, 0) + v_1(y, z), \tag{18}$$

where (5) is assumed to hold. The following result holds [5].

Theorem 3:

Assume that $\|f\|_2$, $\|v_0\|_{\langle 5/2 \rangle}$, $\|v_1\|_{\langle 5/2 \rangle}$ *and* $\|\mathbf{T}_p\|_{\langle 2 \rangle}$ *are sufficiently small. Then there exists a solution of (1)-(3) subject to the given boundary conditions such that* $v \in H^3$, $p \in H^2$, $\mathbf{T} \in H^2$. *Except for a constant in the pressure, this solution is the only one for which* $\|v - (V, 0, 0)\|_3 + \|\mathbf{T}\|_2$ *is small.*

4. Maxwell fluid, supercritical case

If $\rho V^2 > \eta\lambda$, then the system consisting of (11) and (2) is no longer elliptic. It would be wrong to prescribe the boundary conditions of Section 3. This is demonstrated explicitly in [7]. An analysis of the supercritical case is given in [9]. First, we specialize the constitutive relation. In (3), we assume that

$$g_{ij}(\nabla v, \mathbf{T}) = -\frac{\partial v_i}{\partial x_k} T_{kj} - T_{ik}\frac{\partial v_j}{\partial x_k}$$

$$+P_{ik}(\mathbf{T})(\frac{\partial v_k}{\partial x_j} + \frac{\partial v_j}{\partial x_k}) + P_{jk}(\mathbf{T})(\frac{\partial v_k}{\partial x_i} + \frac{\partial v_i}{\partial x_k}) + s_{ij}(\mathbf{T}), \tag{19}$$

where the matrix $\mathbf{P}$ is symmetric. Moreover, $\mathbf{P}$, $\mathbf{s}$ and the first derivatives of $\mathbf{s}$ vanish at $\mathbf{T} = \mathbf{0}$. If one now takes the curl of (11), one obtains a second order hyperbolic system for the vorticity $\zeta = \text{curl } v$ (cf. [2]). This system can be solved when initial conditions for $\zeta(0, y, z)$ and $\frac{\partial \zeta}{\partial x}(0, y, z)$ are prescribed. The velocity field must then be reconstructed from its curl and divergence. However, the condition that ζ must be a curl leads to restrictions on the data which can be prescribed. We refer to [9] and simply state the final result. The following data can be prescribed: Stresses at the inflow boundary according to (17), the tangential part of the curl and its normal derivative at the inflow boundary,

$$\zeta_2(0, y, z) = \zeta_2^0(y, z), \;\; \zeta_3(0, y, z) = \zeta_3^0(y, z),$$

$$\frac{\partial \zeta_2}{\partial x}(0, y, z) = \zeta_2^1(y, z), \;\; \frac{\partial \zeta_3}{\partial x}(0, y, z) = \zeta_3^1(y, z), \tag{20}$$

and the normal velocities at both boundaries,

$$v_1(0, y, z) = a(y, z), \;\; v_1(1, y, z) = b(y, z), \tag{21}$$

where

$$\int_0^L \int_0^M a(y, z)\, dz\, dy = \int_0^L \int_0^M b(y, z)\, dz\, dy. \tag{22}$$

In contrast to the results discussed in Sections 2 and 3, we allow a linear pressure gradient in the y- and z-direction in [9] and we prescribe the mean flow rates

$$\int_0^1 \int_0^L \int_0^M v_2(x, y, z)\, dz\, dy\, dx = \alpha, \;\; \int_0^1 \int_0^L \int_0^M v_3(x, y, z)\, dz\, dy\, dx = \beta. \tag{23}$$

The following theorem holds.

Theorem 4:

Assume that $\|f\|_4$, $\|a\|_{\langle 7/2 \rangle}$, $\|b\|_{\langle 7/2 \rangle}$, $\|\zeta_2^0\|_{\langle 3 \rangle}$, $\|\zeta_3^0\|_{\langle 3 \rangle}$, $\|\zeta_2^1\|_{\langle 2 \rangle}$, $\|\zeta_3^1\|_{\langle 2 \rangle}$, $\|\mathbf{T}_p\|_{\langle 3 \rangle}$, $|\alpha|$ and $|\beta|$ are sufficiently small. Then there is a solution of (1)-(3) which satisfies the given boundary conditions and has the regularity $v \in H^4$, $\mathbf{T} \in H^3$. Moreover, except for a constant in the pressure, this solution is the only one for which $\|v - (V, 0, 0)\|_4$ and $\|\mathbf{T}\|_3$ are small.

Acknowledgement

This research was supported by the National Science Foundation under Grant DMS-8796241.

References

[1] S.R. Burdette, R.C. Armstrong and R.A. Brown, Calculations of viscoelastic flow through an axisymmetric corrugated tube using the explicitly elliptic momentum equation formulation (EEME), *J. Non-Newt. Fluid Mech.*, to appear

[2] D.D. Joseph, M. Renardy and J.C. Saut, Hyperbolicity and change of type in the flow of viscoelastic fluids, *Arch. Rat. Mech. Anal.* **87** (1985), 213-251

[3] M. Luskin, On the classification of some model equations for viscoelasticity, *J. Non-Newt. Fluid Mech.* **16** (1984), 3-11

[4] J.G. Oldroyd, Non-Newtonian effects in steady motion of some idealized elastico-viscous liquids, *Proc. Roy. Soc. London* **A 245** (1958), 278-297

[5] M. Renardy, Inflow boundary conditions for steady flow of viscoelastic fluids with differential constitutive laws, *Rocky Mt. J. Math.* **18** (1988), 445-453

[6] M. Renardy, Recent advances in the mathematical theory of steady flows of viscoelastic fluids, *J. Non-Newt. Fluid Mech.* **29** (1988), 11-24

[7] M. Renardy, Boundary conditions for steady flows of non-Newtonian fluids, in: P.H.T. Uhlherr (ed.), *Proc. Xth Int. Congr. Rheology*, Sydney 1988, Vol. 2, 202-204

[8] M. Renardy, Existence of steady flows of viscoelastic fluids of Jeffreys type with traction boundary conditions, *Diff. Integral Eq.*, to appear

[9] M. Renardy, A well-posed boundary value problem for supercritical flow of viscoelastic fluids of Maxwell type, in: B. Keyfitz and M. Shearer (eds.), *Nonlinear Evolution Equations that Change Type*, IMA Vol. in Math. and its Appl., Springer, to appear

[10] I.M. Rutkevich, The propagation of small perturbations in a viscoelastic fluid, *J. Appl. Math. Mech.* **34** (1970), 35-50

[11] J.S. Ultman and M.M. Denn, Anomalous heat transfer and a wave phenomenon in dilute polymer solutions, *Trans. Soc. Rheol.* **14** (1970), 307-317

For additional references see the Appendix

4. Hyperbolic–Elliptic Problems

THE USE OF VECTORFIELD DYNAMICS IN FORMULATING ADMISSIBILITY CONDITIONS FOR SHOCKS IN CONSERVATION LAWS THAT CHANGE TYPE

Barbara Lee Keyfitz[1]
Department of Mathematics, University of Houston
Houston, Texas 77204-3476, USA

ABSTRACT. Systems of conservation laws which are not of classical hyperbolic type have appeared in modelling complex flows. To formulate admissibility conditions for shocks, we regularize a single shock, between a point where the system is hyperbolic and a point where it is not, by the addition of a higher order term ("viscosity matrix"). This leads to a criterion based on the existence of connecting orbits in a related dynamical system. Local unfoldings of vectorfields are used to analyse the dynamics.

1. INTRODUCTION

This paper summarizes some recent work of the author's on systems of conservation laws that change type. Further details and proofs can be found in [14].

Consider a pair of conservation laws in one space variable (x) and time (t):

$$w_t + h_x \equiv w_t + A(w)w_x = 0 . \tag{1.1}$$

The characteristic speeds λ_i are ordered $\lambda_1 < \lambda_2$ when real. The system is hyperbolic or elliptic according as the discriminant,

$$D(w) = (\mathrm{tr}A)^2 - 4 \det A ,$$

is positive or negative. Change of type means that there are open sets

$$H = \{ w \mid D(w) > 0 \}, \qquad E = \{ w \mid D(w) < 0 \},$$

representing each type, and that their common boundary,

$$B = \{ w \mid D(w) = 0 \} , \tag{1.2}$$

is a smooth curve.

---

[1]Research supported in part by the Air Force Office of Scientific Research, Air Force Systems Command, USAF, under Grant Number AFOSR 86-0088. The U. S. Government is authorized to reproduce and distribute reprints for Governmental purposes notwithstanding any copyright notation thereon.

Equations with this character have been advanced as models for propagation of phase boundaries in two-phase elasticity [9], in fluids of van der Waals type near the critical temperature [22], and in the dynamics of some models for shape-memory alloys or austenitic to martensitic transitions in solids [1]. Change of type of this sort has also been noted in models for pressure-driven, convection-dominated, three-phase flow in porous media [2]. Other experiments in modelling of flows, for example a kinematic model for two-directional traffic flow [3] and an ecological model [8], have also led to systems which change type in this way. An important engineering problem, compressible two-phase flow [24], leads to a larger system of equations that may also contain behavior of this type [16].

Phenomenologically, the flows listed above fall into two classes. In models for the dynamics of phase transitions, the change of type (specifically, the presence of an elliptic region) is coexistent with the presence of a physically unstable range of the order variable (typically density or strain), as manifested by the behavior of the constitutive relation. A typical system is

$$\begin{cases} u_t - v_x = 0 \\ v_t - \sigma(u)_x = 0 \end{cases} . \tag{1.3}$$

This system is a nonlinear wave equation whose characteristic speeds, $\pm\sqrt{\sigma'}$, are real and opposite when $\sigma(u)$ is an increasing function of u, and complex, with real part zero, when $\sigma(u)$ is decreasing. The symmetry of the wave speeds expresses Galilean invariance of the system. We shall refer to systems with this character as of <u>wave equation type</u>. This symmetry is absent in three-phase porous medium flow, where the pressure differential breaks the reflectional symmetry; neither does it occur in the kinematic model of traffic flow nor in typical abstract examples, such as general perturbations of quadratic flux functions, [7]. We find this second class to be <u>generic</u>; all wave-equation-type examples exhibit a degeneracy.

In system (1.1), the question of well-posedness centers on the stability of the initial-value problem. Change of type in a system of conservation laws occurs also in steady transonic flow; the equations are of the form

$$h_x + k_y \equiv A(w)w_x + B(w)w_y = 0 , \tag{1.4}$$

and the characteristic directions, which are the roots of

$$\det |\xi A(w) + \eta B(w)| = 0 ,$$

are real on one side of the sonic line (corresponding to B in (1.2)), and complex on the other. An attempt to formulate admissibility conditions for systems of mixed type based on the model (1.4) was made by Mock [19]; although there is some overlap, the results in [19] do not apply here.

2. CONSERVATION LAW CONSIDERATIONS

In general, solutions to (1.1) exist only in a weak sense: $w(x,t)$ is a measurable function and the pair (w, h) satisfies an integrated form of (1.1); in addition, it is known that w must satisfy some other condition in order for the Cauchy problem to be well-posed [23]. Such conditions are often called _entropy conditions_, by analogy with the manner in which a system like (1.1) is treated in gas dynamics. A number of different formulations of admissibility can be given; in the case of the gas dynamics equations, these all coincide. For systems that change type, such is not the case, and the motivation for the present study, as well as the main application of the result presented here, is to investigate how sensitive such systems might be to different forms of admissibility condition.

The feature we shall consider here is the nature of _shocks_: solutions of (1.1) in the neighborhood of discontinuities. We shall study them by means of _centered shocks_: weak solutions of (1.1) of the form

$$w(x, t) = \begin{cases} w_0 , & x < st \\ w_1 , & x > st \end{cases} \tag{2.1}$$

where w_0 and w_1 are constant states, related to the shock speed s by the _Rankine-Hugoniot_ relation:

$$s[w] - [h] = s(w_1 - w_0) - (h(w_1) - h(w_0)) = 0 . \tag{2.2}$$

Equation (2.2) states that (2.1) is a weak solution of (1.1); for every fixed w_0 the set of values w such that (2.2) is satisfied for some s forms the _wave locus_,

$$W(w_0) = \{ w \mid \exists s \ni s(w - w_0) = h(w) - h(w_0) \} . \tag{2.3}$$

Locally, W is a curve, parameterized by s. Not all points on $W(w_0)$ correspond to allowable shocks, and an entropy condition is required to distinguish those that do. We shall discuss two in this paper:

The _Lax entropy condition_ states

$$\lambda_1(w_0) > s > \lambda_1(w_1)\ ,\qquad \lambda_2(w_0) > s\ ,\qquad \lambda_2(w_1) > s\ , \tag{2.4}$$

for a 1-shock, and

$$\lambda_2(w_0) > s > \lambda_2(w_1)\ ,\qquad \lambda_1(w_0) < s\ ,\qquad \lambda_1(w_1) < s\ , \tag{2.5}$$

for a 2-shock solution to (2.1). This expresses the compressive nature of the shock.

The <u>viscous</u> <u>profile</u> or <u>travelling</u> <u>wave</u> criterion admits as shocks those solutions of (2.1) which are limits as $\varepsilon \to 0$ of self-similar solutions of

$$w_t + h_x = \varepsilon(Mw_x)_x\ , \tag{2.6}$$

$$w = w(\xi) \equiv w\left(\frac{x-st}{\varepsilon}\right)\ ,\qquad w(\xi) \to \begin{cases} w_0\ , & \xi \to -\infty \\ w_1\ , & \xi \to \infty \end{cases} \tag{2.7}$$

This is equivalent to the existence of a connecting orbit from w_0 to w_1 in the vectorfield

$$M\,\frac{dw}{d\xi} = h(w) - sw + c\ , \tag{2.8}$$

where c is the value $sw_0 - h(w_0) = sw_1 - h(w_1)$, from (2.2). Here M, the viscosity matrix, may be constant or variable, singular or nonsingular. The eigenvalues of M must have nonnegative real part; usually M is chosen to be positive definite or semi-definite or even simply the identity matrix (so-called "artificial viscosity") to provide a test for admissibility.

Historically, the viscous profile criterion was introduced, not as an admissibility criterion, but in order to identify the perturbations of parabolic type (2.6) for which one found correct shocks (shocks that satisfied the Lax condition) in the limit. Matrices which admitted shocks that were known to be admissible, and no other shocks, were called "suitable"; studies in this spirit were undertaken by Conley and Smoller, [4], and by Pego and Majda, [18], [20].

3. THE SHOCK LOCUS

We investigate the shock locus when one state is in H and the other in E. The principal result on the shape of the shock locus is given in Theorem 3.1. Proofs of the results in this section can be found in [14].

If $\nabla_w D(w) \neq 0$ on B, then B is a smooth curve. This implies that

$$N(w) = A(w) - \lambda(w)I$$

is a nilpotent matrix of rank one on B ([13]). Denote by r and ℓ its right and left eigenvectors on B. Impose a second nondegeneracy condition

$$\ell^T d^2 h(r, r) > 0 \tag{3.1}$$

on B. If (3.1) holds, then (1.1) is genuinely nonlinear in H, sufficiently close to B, and we prove [14]:

THEOREM 3.1. Let w be a point in B where (3.1) holds. Then for w_0 in a neighborhood $\mathcal{N}$ of w, $W(w_0)$ has the following structure:

if $w_0 \in B$, W forms a cusp opening into H, with axis tangent to $r(w_0)$. The value of s at w_0 is $\lambda(w_0)$, and $ds/d\mu = \mathcal{O}(1/\sqrt{\mu})$ near w_0, where μ is arclength on W.

if $w_0 \in E$, W consists of the point w_0 itself and of a smooth curve lying entirely in H, (and therefore disconnected from w_0); s is monotonic on the curve.

if $w_0 \in H$, W consists of a loop which crosses B, and two segments which leave $\mathcal{N}$ in the opposite direction; thus, W is a curve with a self-intersection; s is monotonic along the entire curve.

We shall refer to the segment between the two self-intersections as the Hugoniot loop. The proof of this theorem uses unfolding theory of steady-state bifurcation problems [5]. A similar theorem was proved earlier by Mock [19], using a different technique.

We now wish to study the vectorfield (2.8) for w_0 in $\mathcal{N}$ and w_1 in $W(w_0)$. Consider a degenerate case: $w_1 = w_0$ and $s = \lambda_i(w_0)$. The local result we seek is found by applying vectorfield unfolding theory [6]. The following theorem is proved in [14]:

THEOREM 3.2. Under assumption (3.1) and an additional nondegeneracy condition

$$\ell^T d^2 h(r, \ell) + r^T d^2 h(r, r) \neq 0 , \tag{3.2}$$

(2.8) at $w_0 = w_1 \in B$, $M = I$ is equivalent (in the sense of vectorfield equivalence) to

$$\begin{aligned} \dot{x} &= y \\ \dot{y} &= ax^2 + bxy \end{aligned} , \quad a, b, \neq 0 . \tag{3.3}$$

Furthermore, varying w_0, w_1, and M in a neighborhood of these values gives a universal unfolding of the vectorfield.

REMARKS. System (3.3) is the Takens-Bogdanov normal form. It has codimension two, in the sense of vectorfield unfoldings, and is discussed in detail by Guckenheimer and Holmes, [6, p365ff]. It is not necessary to vary M to obtain the universal unfolding, but it is useful to recognize that $M \neq I$ is included in the application of this theorem.

The additional nondegeneracy condition (3.2) is necessary to obtain the normal form (3.3), and it is not always satisfied. For example, the nonlinear wave equation, (1.3), is always degenerate. On the other hand, the model equations devised in [11] and [12] to explain some features of a three-phase porous medium flow system, and quadratic models, such as [3], [7] and [8], satisfy this condition, except for special values of the parameters. For this reason, such systems were called generic in the introduction.

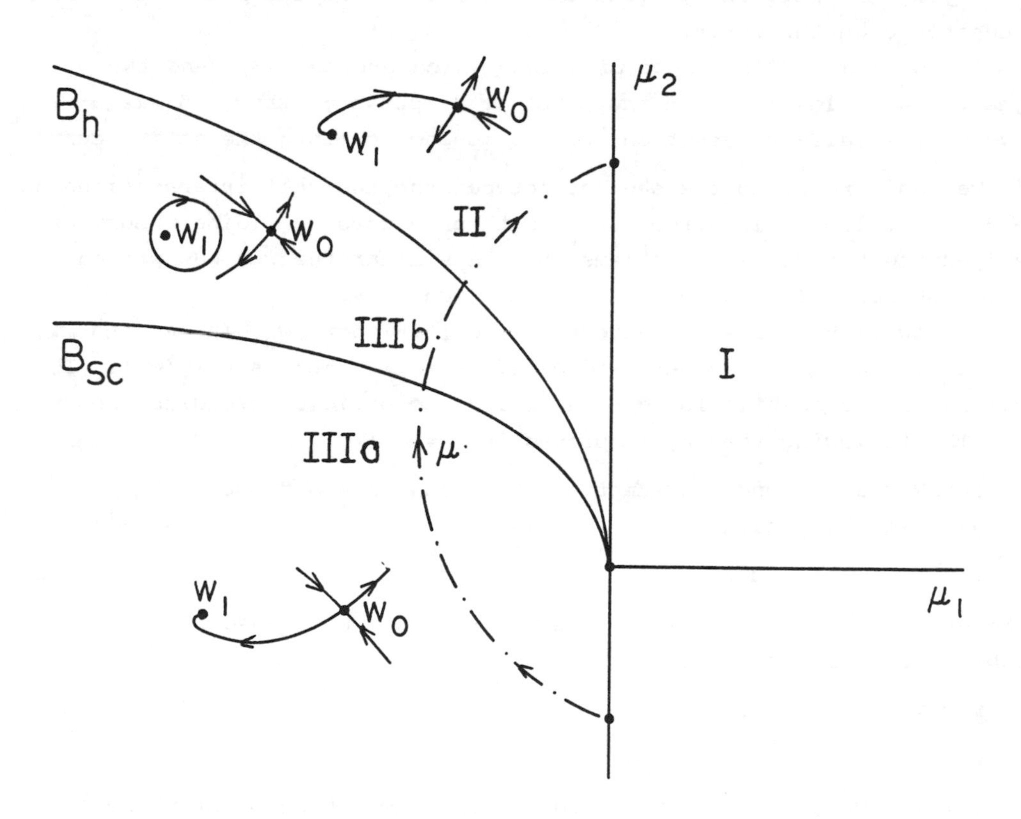

Figure 1

A complete qualitative analysis of the vectorfield (2.8) for w_0 and w_1 in $\mathcal{N}$ is possible. Changing emphasis slightly, assume w_0 to be in H, and w_1 in the Hugoniot loop of w_0; allow w_0 to represent either the right or the left state in (2.1) - that is, either the forward or the backward limit in (2.7). We refer to Figure 1, which follows [6 ,p371], for the case $b > 0$. The coordinates in the Figure are the unfolding parameters. The Hugoniot loop traces a path in this space beginning on the negative μ_2 axis (when $w_1 = w_0$ and $s = \lambda_2$) and ending on the positive μ_2 axis ($w_1 = w_0$ and $s = \lambda_1$), and traversing regions IIIa, IIIb and II in the left half-plane. In region IIIa, there is a connecting orbit: a solution of (2.6), (2.7) with limit (2.1) as $\varepsilon \to 0$. States w_1 near w_0 are classical 2-shocks as long as w_1 remains in H. However, the transition of w_1 into E does not affect the existence or qualitative properties of the orbit. It is true that if $M = I$, then $w_1 \in B$ marks a change of type of the unstable critical point, from a node to a spiral, but this transition occurs at a different value of w_1 if M is perturbed away from the identity.

The other end of the loop, near the positive μ_2 axis, corresponds to the centered shock wave given by

$$w(x, t) = \begin{cases} w_1 , & x < st \\ w_0 , & x > st \end{cases} .$$

This orbit also persists as w_1 crosses B into E.

The curves B_h and B_{sc} enclose a region, IIIb, in which no connecting orbit exists, in either direction. According to the sign of b in (3.3), B_h and B_{sc} are as shown here ($b > 0$) or interchanged. At a point on B_h, a Hopf bifurcation occurs at w_1. For $M = I$, a simple calculation in (2.8) shows that this occurs when

$$s(w_1, w_0) = \mathrm{Re}\ \lambda_1(w_1) = \mathrm{Re}\ \lambda_2(w_1) \tag{3.4}$$

and hence that w_1 is in E at this point. For any other choice of M, the point may be similarly calculated; this is always a local condition.

In IIIb, w_1 is a stable spiral, and an unstable limit cycle separates all orbits containing w_1 from any orbit reaching w_0. Geometrically, this limit cycle grows in size as μ leaves B_h. The curve B_{sc} represents a line of saddle connections: the limit cycle coincides with a homoclinic loop at w_0, which disappears in region IIIa. (This is the "blue sky catastrophe".) The existence of B_{sc} could be inferred from the necessity of continuous transitions of the vectorfield along the Hugoniot loop, but the fact that it is unique, under the nondegeneracy conditions holding for (3.3) and its unfolding, is a consequence of the theory in [6]. Furthermore, it is theoretically quite difficult to find the point on the Hugoniot loop at which $w_1 \in B_{sc}$, since this, unlike (3.4), is not a local condition.

We discuss briefly an interpretation of these conclusions.

Existence of travelling wave orbits for (2.8) can be related to other admissibility criteria. The Lax condition (2.4) or (2.5) is evidently equivalent to the viscous profile criterion for matrices M close to I. (See [18] for the strictly hyperbolic case.)

Theorem 3.1 explicitly excludes system (1.3), which does not satisfy (3.2). For the case $M = I$, it can be verified directly that the curve B_h exists and that along this curve the system is Hamiltonian; there is no region IIIb in this case. Perturbation of this system, say by $M \neq I$, would generally introduce both $b > 0$ and $b < 0$; the normal form for this singularity requires third-order terms. In fact, however, for the models that lead to (1.3), the states in E are usually considered unstable and are of little interest, and an extension of the unfolding technique in another direction might be more useful: for equations of the form (1.3), one can organize the steady-state and vectorfield bifurcations around a singular case: $\sigma' \geq 0$ everywhere, $\sigma' = 0$ at a unique point (a parabolic degeneracy in (1.3), see [10]), with $w_0 = w_1$ located on the degenerate curve. Admissibility for certain phase boundaries could be derived under this criterion. For another, physically motivated, approach to viscous perturbation of (1.3), see [21].

Viscous perturbation of the transonic equations (1.4) is studied in [15] and [17]; there it is shown that, although the steady-state equations are the same, the dynamics is completely different: a natural choice for viscous perturbations gives rise to a dynamical systems unfolding that is essentially one-dimensional (that is, codimension one) and corresponds to the physically reasonable admissibility of all compressible shocks.

5. ACKNOWLEDGEMENTS

I would like to thank the organizers of the GAMM International Conference on Problems Involving Change of Type for the invitation to present these results at the conference honoring Jack Hale. This paper contains a summary of the results in [14].

REFERENCES

1. H. W. Alt, K.-H. Hoffman, M. Niezgódka, and J. Sprekels, "A numerical study of structural phase transitions in shape memory alloys", Preprint No. 90, Inst. for Math., University of Augsburg, 1985.

2. J. B. Bell, J. A. Trangenstein and G. R. Shubin, "Conservation laws of mixed type describing three-phase flow in porous media", SIAM Jour. Appl. Math. 46 (1986), 1000-1023.

3. J. H. Bick and G. F. Newell, "A continuum model for two-directional traffic flow", Quart. Appl. Math. 18 (1960), 191-204.

4. C. C. Conley and J. A. Smoller, "Viscosity matrices for two-dimensional nonlinear hyperbolic systems", Comm. Pure Appl. Math. 23 (1970), 867-884.

5. M. Golubitsky and D. G. Schaeffer, "Singularities and Groups in Bifurcation Theory", Springer-Verlag, New York, 1985.

6. J. Guckenheimer and P. Holmes, "Nonlinear Oscillations, Dynamical Systems, and Bifurcations of Vector Fields", Springer-Verlag, New York, 1983.

7. H. Holden and L. Holden, "On the Riemann problem for a prototype of a mixed type conservation law, II", in Current Progress in Hyperbolic Systems: Riemann Problems and Computations (ed W. B. Lindquist), Contemp. Math. 100, Amer. Math. Soc., Providence, 1989.

8. L. Hsiao and P. de Mottoni, "Existence and uniqueness of Riemann problem for nonlinear system of conservation laws of mixed type", to appear in Trans. Amer. Math. Soc.

9. R. D. James, "The propagation of phase boundaries in elastic bars", Arch. Rat. Mech. Anal. 73 (1980), 125-158.

10. B. L. Keyfitz and H. C. Kranzer, "The Riemann problem for a class of hyperbolic conservation laws exhibiting a parabolic degeneracy", Jour. Diff. Eqns. 47 (1983), 35-65.

11. B. L. Keyfitz, "An analytic model for change of type in three-phase flow", in Numerical Simulation in Oil Recovery, (M. F. Wheeler, ed), IMA Vol 11 (1988), Springer, New York, 149-160.

12. B. L. Keyfitz, "Change of type in three-phase flow: a simple analogue", Jour. Diff. Eqns. 80 (1989), 280-305.

13. B. L. Keyfitz, "A criterion for certain wave structures in systems that change type", in Current Progress in Hyperbolic Systems: Riemann Problems and Computations (ed W. B. Lindquist), Contemp. Math. 100, Amer. Math. Soc., Providence, 1989.

14. B. L. Keyfitz, "Admissibility conditions for shocks in systems that change type", submitted to SIAM Jour. Math. An.

15. B. L. Keyfitz, "Shocks near the sonic line: a comparison between steady and unsteady models for change of type", to appear in Proceedings of IMA workshop on Nonlinear Evolution Equations that Change Type (ed. Keyfitz and Shearer).

16. B. L. Keyfitz, "Change of type in simple models for two-phase flow", University of Houston Mathematics Department research report UH/MD-66, 1989.

17. B. L. Keyfitz and G. G. Warnecke, "The existence of viscous profiles and admissibility for transonic shocks", preprint.

18. A. Majda and R. L. Pego, "Stable viscosity matrices for systems of conservation laws", Jour. Diff. Eqns. 56 (1985), 229-262.

19. M. S. Mock, "Systems of conservation laws of mixed type", Jour. Diff. Eqns. 37 (1980), 70-88.

20. R. L. Pego, "Stable viscosities and shock profiles for systems of conservation laws", Trans. Amer. Math. Soc. 282 (1984), 749-763.

21. M. Shearer, "Admissibility criteria for shock wave solutions of a system of conservation laws of mixed type", Proc. Royal Soc. Edinburgh 93A (1983), 233-244.

22. M. Slemrod, "Admissibility criteria for propagating phase boundaries in a van der Waals fluid", Arch. Rat. Mech. Anal. 81 (1983), 301-315.

23. J. Smoller, "Shock Waves and Reaction-Diffusion Equations", Springer-Verlag, New York, 1983.

24. H. B. Stewart and B. Wendroff, "Two-phase flow: models and methods", Jour. Comp. Phys., 56 (1984), 363-409.

On the Cauchy Problem for the Davey-Stewartson System

Jean-Claude Saut
Université Paris XII and Laboratoire d'Analyse Numérique,
CNRS and Université Paris-Sud, Bâtiment 425, 91405 Orsay (France).

Dedicated to J. Hale on his 60th birthday

Abstract. We present several results obtained with J.M. Ghidaglia on the Cauchy problem for the Davey-Stewartson system : existence of solutions, blow up in finite time.

1. Introduction

The Davey-Stewartson system was introduced in [10] as a model for the evolution of weakly nonlinear packets of water waves that travel predominantly in one direction but in which the wave amplitudes are modulated slowly in both horizontal directions. Hence it is a two dimensional generalization of the cubic monodimensional Schrödinger equation. Djordjevic and Redekopp [11] (see also Ablowitz and Segur [3]) have extended the analysis of [10] by including full gravity, surface tension and depth effects. The model is derived from the Euler equations with free boundary under suitable assumptions (small amplitudes, slowly varying modulations, nearly one dimensional waves and a balance of all three effects). The resulting system for the (complex) amplitude $A(\xi, \eta, \tau)$ and the (real) mean flow velocity potential $\phi(\xi, \eta, \tau)$ (which is defined up to an additive constant) writes in dimensionless form

$$iA_\tau + \lambda A_{\xi\xi} + \mu A_{\eta\eta} = \chi \mid A \mid^2 A + \chi_1 A\phi_\xi \tag{1.1}$$

$$\alpha \, \phi_{\xi\xi} + \phi_{\eta\eta} = -\beta(\mid A \mid^2)_\xi \tag{1.2}$$

The (real) coefficients occuring in (1.1), (1.2) depend on the physical parameters of the problem (fluid depth, wave numbers, group velocity,...). μ, β, χ_1, are positive but χ, λ, α can achieve both signs. In particular $\alpha = \frac{gh - C_g^2}{gh}$ (h = fluid depth, C_g = linear group velocity, g = the constant of gravity) is negative if the effects of surface tension are strong enough (see [11]).

While $\mid A \mid \to 0$ as $\xi^2 + \eta^2 \to +\infty$ is a natural boundary condition for (1.1) in all cases, the appropriate boundary condition for (1.2) will depend dramatically on the sign of α. It will be $\phi \to 0$ as $\xi^2 + \eta^2$ when $\alpha > 0$, and roughly speaking that ϕ vanishes ahead of the support of A, (if A is compactly supported), when $\alpha < 0$. This will of course lead to different formulations of the mathematical problem to be solved.

It is noteworthy to mention that in the shallow water limit of (1.1) (1.2), i.e. when $kh \to 0$ with $a \mid \kappa \mid << (kh)^2$, where $\kappa = (k, \ell)$ is the wave vector and a the characteristic amplitude of the disturbance (and in this case only), the Davey-Stewartson system is of inverse scattering type. In this case, (1.1), (1.2) can be written after rescaling

$$iA_t - \sigma A_{xx} + A_{yy} = \sigma \mid A \mid^2 A + A\phi_x \tag{1.3}$$

$$\sigma\phi_{xx} + \phi_{yy} = -2(\mid A \mid^2)_x \tag{1.4}$$

with $\sigma = \pm 1$.

On the other hand, (1.1)(1.2) reduces in the deep water limit ($kh \to \infty$ where k is the wave number in the $x-$direction) to the nonlinear Schrödinger equation in 2 dimensions

$$iA_\tau + \lambda_\infty A_{\xi\xi} + \mu_\infty A_{\eta\eta} = \chi_\infty \mid A \mid^2 A,$$

where λ_∞ is usually negative, leading to an "hyperbolic" Schrödinger equation.

The system (1.3),(1.4) has be derived independently by Ablowitz-Haberman [2], Morris [18] and Cornille [9]. The motivation was in these works to systematically look for two-dimensional generalizations of the cubic onedimensional Schrödinger equation which were of IST type. The study of (1.3)(1.4) under inverse scattering methods has led recently to very interesting issues : existence of special soliton or lump solutions [5], [1], solution of the Cauchy problem ([6][12] and the bibliography of these papers). However, no results on the Cauchy problem for the general system (1.1), (1.2) seemed to be known so far.

We review here some recent results obtained in a joint work with Jean-Michel Ghidaglia [13][14] on existence and properties of solutions to the Cauchy problem for (1.1), (1.2).

We shall study a scaled version of (1.1), (1.2), namely

$$i\frac{\partial A}{\partial t} + \epsilon_1 \frac{\partial^2 A}{\partial x^2} + \frac{\partial^2 A}{\partial y^2} = \chi \mid A \mid^2 A + A\frac{\partial \phi}{\partial x} \tag{1.5}$$

$$\epsilon_2 \frac{\partial^2 \phi}{\partial x^2} + \frac{\partial^2 \phi}{\partial y^2} = -b\frac{\partial}{\partial x} \mid A \mid^2 \tag{1.6}$$

$$A(x, y, 0) = A_0(x, y) \tag{1.7}$$

Since only the sign of ϵ_1, ϵ_2 will be relevant in our analysis we will suppose to simplify the exposition, that $\epsilon_1, \epsilon_2 = \pm 1$.

For what follows, it will be useful to classify (1.5)(1.6) into elliptic-elliptic, hyperbolic-elliptic, elliptic-hyperbolic, hyperbolic-hyperbolic acording to the sign of (ϵ_1, ϵ_2) : $(+,+), (-,+), (+,-)$ and $(-,-)$.

In all these cases, assuming that A and ϕ are smooth enough and decay suitably at infinity, (1.5) (1.6) admits three interesting integrals [3].

$$M(t) \equiv \int \mid A(t) \mid^2 dxdy = M(0), \ \forall t \tag{1.8}$$

$$E(t) \equiv \int \left[\epsilon_1 \left| \frac{\partial A}{\partial x} \right|^2 + \left| \frac{\partial A}{\partial y} \right|^2 - \frac{1}{2} \left((-\chi) \mid A \mid^4 + \frac{1}{b} \left(\epsilon_2 \left| \frac{\partial \phi}{\partial x} \right|^2 + \left| \frac{\partial \phi}{\partial y} \right|^2 \right) \right) \right] dxdy \qquad (1.9)$$
$$= E(0), \ \forall\, t$$

$$\frac{d^2}{dt^2} \int (\epsilon_1 x^2 + y^2) \mid A \mid^2 dxdy = 8E(0), \ \forall\, t \qquad (1.10)$$

The two conservation laws (1.8)(1.9) are respectively that of mass and energy while (1.10) express the evolution of the moment of inertia.

2. The Cauchy problem in the elliptic-elliptic and hyperbolic-elliptic case.

The first results do not use the conservation of energy.

Theorem 1 *(Existence and uniqueness)*

(i) *Let $A_0 \in L^2(\mathcal{R}^2)$. Then there exists a unique maximal solution (A, ϕ) of (1.5) (1.6) (1.7) on $[0, T^*)$, $T^* > 0$ which is such that*
$A \in C([0,T^*)\ ;\ L^2(\mathcal{R}^2)) \cap L^4((0,t) \times \mathcal{R}^2)$,
$\nabla\phi \in L^2((0,t) \times \mathcal{R}^2)$, $\| A(t) \|_{L^2} = \| A_0 \|_{L^2}$, $0 \le t < T^*$.

(ii) *If A_0 is sufficiently small in $L^2(\mathcal{R}^2)$, $T^* = +\infty$: the solution is global.*

Theorem 2 *(smoothness)*

(i) *Let $A_0 \in H^1(\mathcal{R}^2)$, then the solution (A, ϕ) satisfies moreover*
$A \in C([0,T^*); H^1(\mathcal{R}^2)) \cap C^1([0,t]; H^{-1}(\mathcal{R}^2))$
$\nabla A \in L^4((0,t) \times \mathcal{R}^2)$, $\nabla\phi \in C([0,t]; L^p(\mathcal{R}^2))$ *and*
$\nabla^2\phi \in L^4(0,t; L^q(\mathcal{R}^2))$, *for every* $t \in [0,T^*)$, $p \in [2,+\infty)$, $q \in [2,4)$.

(ii) *If $A_0 \in H^2(\mathcal{R}^2)$, then*
$A \in C([0,T^*); H^2(\mathcal{R}^2)) \cap C^1([0,t]; L^2(\mathcal{R}^2))$
$\nabla\phi \in C([0,T^*); H^2(\mathcal{R}^2))$
$A \in L^2(0,t; H^{5/2}_{\text{loc}}(\mathcal{R}^2))$, $\nabla^2\phi \in L^1(0,t; H^{3/2}_{\text{loc}}(\mathcal{R}^2))$ *for every* $\mathbf{t} \in [\mathbf{0}, \mathbf{T}^*)$.

Theorem 3 *(Continuous dependence).*

Let $I = [0,T]$, for some $T > 0$. The map $A_0 \to (A, \nabla\phi)$ is continuous from $H^1(\mathcal{R}^2)$ into $C(I; H^1(\mathcal{R}^2)) \times C(I; L^p(\mathcal{R}^2))$, $2 < p < +\infty$. More precisely, let $A \in C(I; H^1(\mathcal{R}^2)$, $\nabla\phi \in C(I; L^p(\mathcal{R}^2))$ be a solution of (1.5) (1.6) (1.7) with $A(0) = A_0$. Let $A_{0n} \to A_0$ in $H^1(\mathcal{R}^2)$ as $n \to +\infty$. Then the solution (A_n, ϕ_n) with $A_n(0) = A_{0n}$ exists on the interval I if n is sufficiently large, and $(A_n, \nabla\phi_n) \to (A, \nabla\phi)$ in $C(I; H^1(\mathcal{R}^2)) \times C(I; L^p(\mathcal{R}^2))$, $2 < p < +\infty$.

Remarks 1.

1. When $b = 0$ and $\epsilon_1 = -1$, (1.5) reduces to an "hyperbolic" Schrödinger equation

$$i\frac{\partial A}{\partial t} - \frac{\partial^2 A}{\partial x^2} + \frac{\partial^2 A}{\partial y^2} = \chi \mid A \mid^2 A, \tag{1.11}$$

to which our results apply. Further properties of Schrödinger equations like (1.11) are investigated in [15].

2. In the context of Theorem 1, the energy (1.9) is in general not defined and one cannot expect its conservation. On the other hand, when $A_0 \in H^1(\mathcal{R}^2)$ (Theorem 2), the energy is well defined and conserved. Furthermore, when the mass M and E bound a priori the H^1 norm of A (i.e. $\epsilon_1 = 1$, $\epsilon_2 = 1$, $\chi \geq \max(b^{-1}, 0)$, the solution constructed in Theorem 2 is global.

3. Using inverse scattering techniques, Beals and Coifman have recently proved [6] global existence in $S(\mathcal{R}^2)$ for the Cauchy problem corresponding to (1.3), (1.4) with $\sigma = +1$.

Indications on the proofs. We will just give some comments on the proofs (see [14] for details).

(i) We use (1.6) to express $\nabla\phi$ in terms of $\mid A \mid^2$, by performing successively two Riesz transforms. Hence

$$\| \nabla\phi \|_{L^p(\mathcal{R}^2)} \leq Cp \| A \|^2_{L^{2p}(\mathcal{R}^2)}, \; 1 < p < +\infty \tag{1.12}$$

This allows to transform (1.5) (1.6) into a nonlocal nonlinear Schrödinger equation with will be studied on its integral formulation.

$$A(t) = S(t)A_0 - i\int_0^t S(t-s)[\chi \mid A \mid^2 A + AE(A)](s)ds \tag{1.13}$$

where $S(t)$ denotes the group associated to the linear Schrödinger equation and $E(A) = \phi_x$.

The crux of the proof of Theorems 1 and 2 is to use contraction type arguments in spaces such as $L^4(0, T; L^4(\mathcal{R}^2))$, similar to those used by Cazenave-Weissler [7], Kato [17] in the context of the nonlinear "elliptic" Schrödinger equation. This method needs $L^p - L^q$ estimates for the propagator of the linear Schrödinger equation. They are classical for the usual "elliptic" Schrödinger equation, but are also valid in the more general case of the propagator associated to the linear equation.

$$i\frac{\partial u}{\partial t} + \sum_{i,j=1}^{n} a_{ij}\frac{\partial^2 u}{\partial x_i \partial x_j} = 0 \text{ in } \mathcal{R}^n \tag{1.14}$$

where the real constants a_{ij} are such that the matrix (a_{ij}) is invertible.

(iii) The proof of the smoothness properties follows the same lines. The second part of (ii) in Theorem 2 results from a local smoothing effect for the linear Schrödinger equation [8], which is still valid for the more general equation (1.14).

So far we have not used the conserved quantities (1.9) (1.10) (except in Remarks 1,2). We will now make use of (1.9) (1.10) to obtain blow-up results in the elliptic-elliptic case, for large initial data.

To begin with we want to justify (1.10), i.e. to prove that the Cauchy problem for (1.5) (1.6) (1.7) is well posed in a weighted Sobolev space.

Theorem 4.
Let $A_0 \in \Sigma = \{v \in H^1(\mathcal{R}^2),\ (x^2+y^2) \mid v \mid^2 \in L^1(\mathcal{R}^2)\}$. *The solution* (A,ϕ) *of* (1.5), (1.6), (1.7) *satisfies* (1.10) *and*

$$(x^2+y^2) \mid A \mid^2 \in C([0,T^*);L^1(\mathcal{R}^2)) \cap L^2((0,T)\times\mathcal{R}^2),\ \forall\, t \in [0,T^*)$$

Idea of the proof. If $A_0 \in H^1(\mathcal{R}^2)$ we know by Theorem 2 that the maximal solution (A,ϕ) on the inverval $[0,T^*)$ given by Theorem 1 is such that $A(t)$ belongs to $H^1(\mathcal{R}^2)$ for $0 \le t < T^*$. Theorem 4 will follow from a continuation argument which says that if $A_0 \in \Sigma$ then $A(t)$ belongs to Σ as long as $A(t) \in H^1(\mathcal{R}^2)$. Use will be made of a commutation result for the linear Schrödinger equation which we state in the context of the general equation (1.14).

Lemma 1. *Let* $A = (a_{ij})$ *be a nonsingular* $n \times n$ *symmetric matrix and let* $B = A^{-1}$. *The linear operator* $J = (J_1,\dots,J_n)$ *given by*

$$J_k v = e^{i\psi(x)/4t}(2it)\frac{\partial}{\partial x_k}(e^{-i\psi(x)/4t}v)$$

where

$$\psi(x) = \sum_{i,j=1}^{n} b_{ij}x_i x_j,\quad B = (b_{ij})$$

commutes with the general linear Schrödinger operator

$$L = i\frac{\partial}{\partial t} + a_{ij}\frac{\partial^2}{\partial x_i \partial x_j}$$

(For the classical Schrödinger operator, $a_{ij} = \delta_{ij}$, this is just the conformal invariance property. See Ginibre and Velo [16]).

As we mentioned in Remarks 1,2, when $\epsilon_1 = \epsilon_2 = 1$ and $\chi \ge \mathrm{Max}(1/b,0)$ the solution (A,ϕ) is global. Let us assume now that $\chi < \mathrm{Max}(b^{-1},0)$. It is then possible to find $A_0 \in \Sigma$ such that $E(0) < 0$. One can proceed for instance in the following way. Let A_0 be the gaussian $A_0(x,y)$ $= \lambda \exp\,[-(\beta^{-2}x^2+\gamma^{-2}y^2)]$ where λ,β,γ are real and positive. The energy E (see (1.9)) writes $a\lambda^2 - J\lambda^4$ where the constant $J = J(A_0)$ has the sign of $-\chi+b^{-1}\gamma/(\beta+\gamma)$. Hence, when $-\chi > 0$, or $-\chi+b^{-1} > 0$ (i.e. $\chi < \mathrm{Max}(b^{-1},0)$), one has $J > 0$ by a suitable choice of γ and β. Then $E(A_0)$ can be made <0 for λ sufficiently large. Note that when $\chi > 0$, it is the contribution of ϕ which makes E negative. In view of (1.10) it is then obvious that the solution (A,ϕ) can only exist on a finite time.

Corollary. Under the hypothesis of Theorem 4, when $\epsilon_1 = +1$ and $E(A_0) < 0$ (see above), the solution (A, ϕ) blows up in finite time, $T^* < \infty$.

Remarks 2.
1. This result has been given formally by Ablowitz and Segur [3].
2. The argument given above cannot clearly be applied when $\epsilon_1 = -1$. The formation of singularities in the hyperbolic-elliptic case (and hence for the hyperbolic Schrödinger equation (1.11)) is an open problem.

3. The Cauchy problem in the elliptic-hyperbolic ase.

It will be assumed in this section that $\epsilon_1 = 1$, $\epsilon_2 = -1$.

Theorem 5. Let $A_0 \in H^1(\mathcal{R}^2)$ be sufficiently small in L^2. Then there exists a solution (A, ϕ) in the sense of distribution of (1.5) (1.6) (1.7) which satisfies

$$A \in L^\infty(0,\infty; H^1(\mathcal{R}^2)), \quad \phi \in L^\infty((0,\infty) \times \mathcal{R}^2).$$

$$M(t) = M(0) \quad \forall\, t \in [0,\infty)$$

Indication on the proof.
(i) In order to express ϕ in function of A, we study the following problem. Given $f \in L^1(\mathcal{R}^2)$, find ϕ solution of the wave equation

$$\frac{\partial^2 \phi}{\partial x^2} - \frac{\partial^2 \phi}{\partial y^2} = f \tag{3.1}$$

which satisfies a radiation condition at infinity : when f is compactly supported, ϕ tends to zero ahead of the support of f. This amounts to saying in characteristics variables $\xi_1 = x + y$, $\xi_2 = x - y$ that

$$\lim_{x+y \to +\infty} \phi(x,y) = \lim_{x-y \to +\infty} \phi(x,y) = 0 \tag{3.2}$$

Such a condition is dictated by physical considerations (when $f = |A|_x^2$, [3], [4]). We state

Proposition 1. *Let $f \in L^1(\mathcal{R}^2)$ and K be the kernel*

$$K(x_1, y_1, x_2, y_2) = H(x_2 + y_2 - x_1 + y_1) H(x_2 - y_2 + y_1 - x_1)/2$$

where H is the Heaviside function. Then, the function

$$\phi(x,y) = \mathcal{H} f(x,y) = \int_{\mathcal{R}^2} K(x, y, x_1, y_1) f(x_1, y_1)\, dx_1 dy_1 \tag{3.3}$$

is continuous and is a solution of (3.1) (3.2). *Moreover*

$$\sup_{(x,y) \in \mathcal{R}^2} |\phi(x,y)| \leq \frac{1}{2} \int_{\mathcal{R}^2} |f(x,y)|\, dx dy \tag{3.4}$$

$$\int_{\mathcal{R}^2}\left|\left(\frac{\partial\phi}{\partial x}\right)^2-\left(\frac{\partial\phi}{\partial y}\right)^2\right|dxdy\leq\left(\int_{\mathcal{R}^2}|f(x,y)|\,dxdy\right)^2 \tag{3.5}$$

Remarks 3.
1. In general $\nabla\phi$ does not belong to $L^2(\mathcal{R}^2)$, even if $f\in\mathcal{D}(\mathcal{R}^2)$.
2. The estimate (3.4) will give a sense to the energy E.

(ii) Construction of approximate solutions and a priori estimates. We consider the regularized problem

$$\begin{cases} i\dfrac{\partial A^\epsilon}{\partial t}+\Delta A^\epsilon+i\epsilon\Delta^2\dfrac{\partial A^\epsilon}{\partial t}=\chi\mid A^\epsilon\mid^2 A^\epsilon+A^\epsilon\dfrac{\partial\phi^\epsilon}{\partial x}, \\ \dfrac{\partial^2\phi^\epsilon}{\partial x^2}-\dfrac{\partial^2\phi^\epsilon}{\partial y^2}=-b\dfrac{\partial}{\partial x}(\mid A^\epsilon\mid^2) \\ A^\epsilon(\cdot,0)=A_0^\epsilon\in H^2(\mathcal{R}^2)\text{ with }A_0^\epsilon\to A_0\text{ in }H^1(\mathcal{R}^2). \end{cases} \tag{3.6}$$

The existence of a unique global solution of (3.6) is proved with Segal's Theorem [19]. Moreover, one obtains the following a priori estimates, provided $\| A_0\|_{L^2}$ is sufficiently small,

$$\begin{aligned} \| A^\epsilon\|_{L^\infty(\mathcal{R}_+;H^1(\mathcal{R}^2))}&\leq C \\ \|\phi^\epsilon\|_{L^\infty(\mathcal{R}_+\times\mathcal{R}^2)}&\leq C \end{aligned} \tag{3.7}$$

where C is independent of ϵ.

(iii) Passing to the limit. Thanks to (3.7) one can extract a subsequence still denotes $(A^\epsilon,\phi^\epsilon)$ such that

$$\begin{aligned} &A^\epsilon\to A\ \text{ in }L^\infty(0,\infty;H^1(\mathcal{R}^2))\text{ weak}^* \\ &A^\epsilon\to A\ \text{ in }L^p_{loc}((0,\infty)\times\mathcal{R}^2)\text{ strongly }p\geq 2 \\ &\phi^\epsilon\to\psi\ \text{ in }L^\infty((0,\infty)\times\mathcal{R}^2))\text{ weak}^* \end{aligned}$$

It is easy to check that (A,ψ) satisfy the equation (in the sense of distributions)

$$\begin{aligned} iA_t+\Delta A&=\chi\mid A\mid^2 A+bA\psi_x \\ A(0)&=A_0. \end{aligned} \tag{3.8}$$

To prove that $\psi=\mathcal{H}(\mid A\mid^2_x)$, we observe first that

$$\lim_{\epsilon\to 0}\int_0^T\int_{\mathcal{R}^2}\mid A^\epsilon-A\mid^2 dxdydt=0,\ \forall\, T>0$$

and one exploits the representation (3.3).

4. Final comments.

1. The hyperbolic-hyperbolic case is totally open. However it does not seem to be relevant to the theory of surface water waves (see [3][11]).

2. Some special solutions (solitons, lumps) have been recently found analytically, in the special cases of system (1.3) (1.4) ([1] [5] [12]). Use of inverse scattering is here a crucial point. We postpone to a forthcoming paper the study of soliton solutions of the general system (1.1), (1.2).

References

1 M.J. Ablowitz, A.S. Fokas, On the inverse scattering transform of multidimensional nonlinear equations related to first-order systems in the plane, *J. Math. Phys., 25(8)*, (1984), 2494-2505.

2 M.J. Ablowitz, R. Haberman, Nonlinear evolution equations, two and three dimensions, *Phys. Rev. Lett. 35,* (1975), 1185-1188.

3 M.J. Ablowitz, H. Segur, On the evolution of packets of water waves, *J. Fluid Mech. 92(4),* 1979, 691-715.

4 M.J. Ablowitz, H. Segur, *Solitons and the Inverse Scattering Transform,* S.I.A.M., Philadelphia, 1981.

5 D. Anker, N.C. Freeman, On the soliton solutions of the Davey-Stewartson equation for long waves, *Proc. R. Soc. London, A 360,* (1978), 529-540.

6 R. Beals, R.R. Coifman, The spectral problem for the Davey-Stewartson and Ishimori hierarchies, preprint, 1988.

7 T. Cazenave, F.B. Weissler, Some remarks on the nonlinear Schrödinger equation in the critical case, preprint, 1987.

8 P. Constantin, J.C. Saut, Local smoothing properties of dispersive equations, *Journal of A.M.S. 1,2,* (1988), 413-433.

9 H. Cornille, Solutions of the generalized nonlinear Schrödinger equation in two spatial dimensions, *J. Math. Phys. 20(1),* (1979), 199-209.

10 A. Davey, K. Stewartson, On three-dimensional packets of surface waves, *Proc. R. Soc. London A 338,* (1974), 101-110.

11 V.D. Djordjevic, L.G. Redekopp, On two dimensional packets of capillary-gravity waves, *J. Fluid Mech. 79(4),* (1979), 703-714.

12 A.S. Fokas, P.M. Santini, Solitons in multidimensions, preprint, 1988.

13 J.M. Ghidaglia, J.C. Saut, Sur le problème de Cauchy pour les équations de Davey-Stewartson, *C.R. Acad. Sci. Paris,* 1988.

14 J.M. Ghidaglia, J.C. Saut, On the Cauchy problem for the Davey-Stewartson equations, to appear.

15 J.M. Ghidaglia, J.C. Saut, Nonelliptic Schrödinger equations, in preparation.

16 J. Ginibre, G. Velo, On a class of nonlinear Schrödinger equations, *J. Funct. Anal., 32* (1979), 1-71.

17 T. Kato, On nonlinear Schrödinger equations, *Ann. Inst. Henri Poincaré, Physique Théorique 46,* (1987), 113-119.

18 H.C. Morris, Prolongation structures and nonlinear evolution equations in two spatial dimensions. II A generalized nonlinear Schrödinger equations, *J. Math. Phys., 18(2),* (1977), 285-288.

19 I. Segal, Nonlinear Semigroups, *Annals of Math., 78,* (1963), 339-364.

Inverse Scattering and Factorization Theory

D.H. Sattinger[1]

School of Mathematics
University of Minnesota

Abstract: The analogy between the inverse scattering problem for the Schrödinger operator and an infinite dimensional factorization problem is discussed. The τ function introduced by Hirota is obtained as the Fredholm determinant of the "lower minors" of the scattering operator. The classical Darboux transformation is presented in the context of the dressing method. A brief account of the "dressing method" for the KdV and Kadomtsev-Petviashvili hierarchies is given.

1. Scattering operators.

The scattering theory of the wave equation in an exterior domain in R^3

$$u_{tt} - \Delta u = 0, \qquad x \in \Omega,$$

where Ω is the exterior of some bounded object in R^3, is given in [13]. Let U_0 (t) be the propagator for the free wave equation; i.e.$\{u,u_t\}= U_0(t)f$ is the solution of the wave equation in all space with initial data $f=\{f_1,f_2\}$ (the initial data is given by u and u_t at some initial instant of time). Let U(t) be the propagator for the wave equation in Ω. The wave operators $W_\pm$ are defined by

$$W_\pm f = \lim_{t\to\pm\infty} U(-t)U_0(t)f$$

and the scattering operator by $S = W_+^{-1}W_-$. Solutions of finite energy tend to a solution of the free wave equation as $|t| \to\infty$. The scattering operator S transforms the solution at $-\infty$ to that at $+\infty$. In this case, it is unitary, since energy is conserved. Furthermore, $[S, U_0]=0$, and $U(t)W_\pm = W_\pm U_0(t)$.

Zakharov and Shabat [22] have developed a theory of inverse scattering for the Schrödinger operator on the line based on factorization theory. (Cf. Melin [15] for a careful

---

AMS (MOS): 35Q20, 34A55

[1]This research was supported in part by NSF grant DMS-87-02578

analysis of the factorization approach to inverse scattering) The so-called "dressing method", as it applies to the theory of inverse scattering theory of the Schrödinger equation in one dimension, runs as follows: Let $D=\partial/\partial x$, L be the Schrödinger operator D^2+u, and let $K_{\pm}$ be Volterra integral operators

$$K_+\psi(x) = \int_x^\infty K_+(x,t)\psi(t)\,dt \qquad\qquad K_-\psi(x) = \int_{-\infty}^x K_-(x,t)\psi(t)\,dt$$

We search for kernels $K_{\pm}$ such that $L(1+K_{\pm}) = (1+K_{\pm})D^2$. This operator identity leads to equations for the kernels $K_{\pm}$. For example, for K_+ we get the following equation, which must hold for all smooth, integrable functions $\psi(x)$:

$$\int_x^\infty (K_{xx}-K_{yy}-u(x)K(x,y))\psi(y)\,dy \;+\; (u(x) - 2\frac{d}{dx}K(x,x))\psi(x) = 0$$

This identity implies that the kernels $K_{\pm}$ satisfy the hyperbolic equation and characteristic boundary condition

$$K_{xx} - K_{yy} - u(x)K(x,y) = 0 \text{ in } x<y, \qquad \text{and } u(x) = 2\frac{d}{dx}K(x,x)$$

This procedure, on which there is an extensive literature, is known as the "dressing method" in the Russian literature [22]; the intertwining operators $1+K_{\pm}$ are sometimes called "transmutation operators" [6, 13].

Putting

$$\psi(x,k) = (1+K_+)e^{ikx} = e^{ikx} + \int_x^\infty K(x,t)e^{ikt}dt$$

we find that ψ is analytic in Im $k \geq 0$ and satisfies the time independent scattering problem

$$(L+k^2)\psi= 0, \qquad \psi\sim e^{ikx} \text{ as } x \to\infty$$

The wave function ψ and the kernel K are related by the Fourier transform. In fact,

$$K_+(x,t) = \frac{1}{2\pi}\int_{-\infty}^\infty (\psi(x,k) - e^{ikx})e^{-ikt}dk$$

For x>t we close the contour of this Fourier integral in the upper half plane and find that $K_+ = 0$ for t<x. Thus, the dressing method transforms the time independent scattering problem for ψ to the time dependent one for K. The analyticity in the wave functions of L transforms into causality conditions on the kernels K_+ and K_-, which are solutions of the hyperbolic equation for radiation coming from $-\infty$ and $+\infty$ respectively.

Given two operators $1+K_\pm$ that both intertwine D^2 with L, let the operator F be formally defined by $(1+F) = (1+K_+)^{-1}(1+K_-)$. Then $L(1+K_-) = L(1+K_+)(1+F) = (1+K_+)D^2(1+F)$ on the one hand, and $L(1+K_-)= (1+K_-)D^2 = (1+K_+)(1+F)D^2$ on the other. Assuming $(1+K_+)$ is invertible, we find that $[D^2, F] = 0$. Denoting the kernel of F by f, we get the free wave equation $f_{xx} - f_{tt}=0$. The general solution of this equation is, of course, $h(x+t) + g(x-t)$; but, for reasons that, as far as I know, do not follow from the preceeding argument alone, it turns out that only the solution $h(x+t)$ is needed. We denote it by $f(x+t)$. A derivation is given below.

In analogy with the scattering theory of the wave equation, the operator (1+F) may be called the scattering operator. The intertwining operators $(1+K_\pm)$ are analogs of the wave operators $W_\pm$. They intertwine the "bare" or free space operator with the perturbed operator.

It is clear that if $L(1+K_+) = (1+K_+)D^2$, then L is a differential operator plus possibly an upper Volterra integral operator. On the other hand, if $L(1+K_-) = (1+K_-)D^2$, then L is a differential operator plus possibly a lower Volterra integral operator. Zakharov and Shabat proved that $[1+F, D^2] = 0$ iff $(1+K_\pm)$ both dress D^2 to the same operator L. Furthermore, in this case, L is a purely differential operator.

2 The Gel'fand-Levitan-Marchenko equation.

The relationship $(1+K_+)(1+F) = 1+K_-$ may be written

$$K_+(x,y) + f(x+y) + \int_x^\infty K_+(x,s)f(s+y)\,ds = K_-(x,y)$$

Since $K_-(x,y) = 0$ for y>x, we get the Gel'fand-Levitan-Marchenko (GLM) integral equation

$$K_+(x,y) + f(x+y) + \int_x^\infty K_+(x,s)f(s+y)\,ds = 0 \qquad y>x$$

Under reasonable decay conditions on f, the integral operator F is of Hilbert-Schmidt type. The kernel f is related to the scattering data of the time independent equation by

$$f(s) = \sum_{j=1}^{N} c_j e^{-\omega_j s} + \frac{1}{2\pi}\int_{-\infty}^{\infty} r(k)e^{iks}\,dk$$

where $-\omega_j^2$ are the discrete eigenvalues of L, r(k) is the reflection coefficient and

$$c_j = \frac{1}{\int_{-\infty}^{\infty} \psi(x,i\omega_j)^2\, dx}$$

where $\psi(x,i\omega_j)$ is the eigenfunction of L with eigenvalue $i\omega_j$.

The derivation of the GLM equation for inverse scattering is by now well known. (cf. [1, 4, 8, 16]). We give here a simplified derivation which is directly related to the factorization $(1+F) = (1+K_+)^{-1}(1+K_-)$. Let $\phi_+(x,k)$ and $\psi_+(x,k)$ be solutions of $(L+K^2)\psi = 0$ which are analytic in Im k >0 and satisfy the asymptotic conditions $\phi_+(x,k) \sim e^{-ikx}$ as $x \to -\infty$ and $\psi_+(x,k) \sim e^{ikx}$ as $x \to +\infty$. Similarly, there exist wave functions $\phi_-(x,k)$ and $\psi_-(x,k)$, analytic in the lower half k - plane, with corresponding asymptotic conditions. As $x \to +\infty$, ϕ_+ is asymptotic (for real k) to $a(k)e^{-ikx} + b_+(k)e^{ikx}$. Across the real k axis the wave functions satisfy the jump conditions

$$\phi_+ = a(k)\,\psi_- + b_+(k)\,\psi_+ \qquad\qquad \phi_- = b_+(k)\,\psi_- + a(k)\,\psi_+$$

It is easy to prove that $2ika(k) = W[\phi_+(x,k), \psi_+(x,k)] = \phi_+\psi_+' - \phi_+'\psi_+$, the Wronskian of the wave functions ϕ_+ and ψ_+. The reflection coefficient is $r(k) = b_+(k)/a(k)$.

Define

$$\Phi(x,k) = \begin{cases} \dfrac{\phi_+(x,k)}{a(k)} - e^{-ikx} & \text{Im } k > 0 \\[2ex] \psi_-(x,k) - e^{-ikx} & \text{Im } k < 0 \end{cases}$$

and let K_γ be the Fourier transform

$$K_\gamma(x,y) = \frac{1}{2\pi}\int_{-\infty}^{\infty} \Phi(x, \xi+i\gamma)\, e^{i(\xi+i\gamma)y} d\xi$$

Here γ is real; and K_γ is the Fourier transform of Φ along the line Im z = γ. K_γ jumps as the contour crosses the singularities of Φ. These consist of poles at the zeroes of a(k) and a jump across γ=0. Let the zeroes of a be $i\omega_j$, j = 1 . . . N, with the maximum being ω_N. For $\gamma > \omega_N$, we can close the contour in the upper half plane, and $K_\gamma = 0$ for x<y. When γ<0 we can close the contour in the lower half plane, and $K_\gamma = 0$ for x>y.

In fact, for γ<0, K_γ coincides with K_+ which we introduced above, and $\psi_+ =$

$(1+K_+)e^{ikx}$. For $\gamma>\omega_N$, K_γ coincides with K_-. However, due to the presence of solitons, $K_-(x,y)$ grows exponentially as $y\to-\infty$, and we have $\phi_-(x,k)/a(k) = (1+K_-)e^{-ikx}$ for Im $k > \omega_N$.

The GLM equation is obtained by comparing the kernels K_- and K_+. Let us compute the contribution to (1+F) from the jump across the real axis. We denote by K_0 the Fourier transform computed along a contour that lies in $0<\mathrm{Re}\, z<\omega_1$. We have

$$K_0(x,y) = \frac{1}{2\pi}\int_{-\infty}^{\infty} e^{i\xi y}\left(\frac{\phi_+(x,\xi)}{a(\xi)} - e^{-i\xi x}\right) d\xi$$

$$= \frac{1}{2\pi}\int_{-\infty}^{\infty} e^{i\xi y}(\psi_-(x,\xi) - e^{-i\xi x} + r(\xi)\,\psi_+(x,\xi))\, d\xi$$

$$= K_+(x,y) + \frac{1}{2\pi}\int_{-\infty}^{\infty} e^{i\xi y} r(\xi)\left(e^{i\xi x} + \int_x^{\infty} K_+(x,s)e^{i\xi s} ds\right)d\xi$$

$$= K_+(x,y) + f_1(x+y) + \int_x^{\infty} K_+(x,s)\, f_1(s+y)\, ds$$

where

$$f_1(s) = \frac{1}{2\pi}\int_{-\infty}^{\infty} e^{i\xi s} r(\xi) d\xi$$

Similarly, as γ crosses the zeroes of a(k), we pick up the discrete terms in the kernel f.

The GLM equation may be considered as an infinite dimensional generalization of the problem of factoring a matrix F into the product of upper and lower triangular matrices: UF=L, where U is an upper and L is a lower triangular matrix. A sufficient condition that ensures that this factorization may be performed is that the lower minors of F do not vanish. That is, let $D_k=\det^-_k F$ be the determinant of the lower k×k minor of F. Then the factorization can be performed so long as none of the D_k vanish.

In R^n the finite dimensional subspaces $V_k = \{\psi|\ \psi_j=0 \text{ for } j>k\}$ are invariant under the action of the group of upper triangular matrices. Similarly, on the line, the subspaces $V_a = \{\psi|\ \psi = 0 \text{ for } x > a\}$ are invariant under the group of Volterra operators $\{1+K_+\}$. The infinite dimensional analog of the lower minors of the matrix F is the Fredholm determinant of the truncated scattering operator, $\det(1+F_x)$, where

$$F_x\psi(x) = \int_x^{\infty} f(x+y)\psi(y)dy .$$

Let us define $\tau(x) = \det(1+F_x)$, where by the determinant we mean the Fredholm determinant. (The reason for the notation τ will be explained below.) This exists if f(s) decays sufficiently rapidly as $s \to +\infty$, for example [18] if

$$\sup_{a<s<\infty} |f(s)|\,(1+|s|^{\nu}) < \infty$$

for $\nu>1$. The Fredholm determinant is given as an infinite series

$$\tau(x) = \sum_{n=0}^{\infty} \frac{1}{n!} \int_x^{\infty}\int_x^{\infty} \cdots \int_x^{\infty} f\begin{bmatrix} y_1 \ldots y_n \\ y_1 \ldots y_n \end{bmatrix} dy_1 \ldots dy_n$$

where

$$f\begin{bmatrix} y_1 \ldots y_n \\ z_1 \ldots z_n \end{bmatrix} = \det \|f(y_j+y_k)\|$$

The solution $K_+(x,y)$ of the GLM equation can be expressed in terms of the Fredholm minor D(x,y) as

$$K_+(x,y) = \frac{D(x,y)}{\tau(x)}$$

where

$$D(x,y) = -\sum_{n=0}^{\infty} \frac{1}{n!} \int_x^{\infty}\int_x^{\infty} \cdots \int_x^{\infty} f\begin{bmatrix} x, y_1 \ldots y_n \\ y, y_1 \ldots y_n \end{bmatrix} dy_1 \ldots dy_n$$

In the course of solving the inverse scattering problem by the GLM equation the potential u is recovered from the boundary condition $u(x) = 2\, d/dx\, K(x,x)$. The kernel K(x,x) can in turn be recovered from the Fredholm determinant by a theorem due to Bargmann (cf. Dyson,[9])

$$K_+(x,x) = \frac{d}{dx} \log \tau(x)$$

This result is obtained by differentiating the series for the Fredholm determinant with respect to x and showing that $D(x,x) = \partial(\tau)/\partial x$.

Hirota (cf. his review article [11]) introduced the transformation $u(x)=2(\log \tau(x))_{xx}$ in his construction of multisoliton solutions of the KdV equation ($u_t=(u_{xxx} + uu_x)/4$) by the direct method [11]. He showed that τ satisfies a certain *bilinear* equation. The connection between Hirota's direct method and the inverse scattering method via the determinant of the scattering operator was observed and developed by Oishi [17] and Pöppe [18] for the Kdv equation and also the sine-Gordon equation. Pöppe showed that $\tau = \det(1+F_x)$ satisfies Hirota's bilinear equation for the KdV and sine-Gordon equations. There are deep connections between the theory of the τ function and soliton equations. Sato [21] explained the τ function as Plücker coordinates on an infinite dimensional Grassmannian manifold. (Cf. also Date *et al.)*

The Schrödinger operator L is an isospectral operator for the KdV equation: $u_t=(u_{xxx} + uu_x)/4$. As u evolves according to the (nonlinear) KdV equation, the scattering data of L evolves linearly. This is the basis of the well-known inverse scattering method for completely integrable systems: one solves for the flow u(x,t) by letting the scattering data evolve linearly, and then solving the inverse scattering problem at time t. The transformation to the scattering data thus linearizes the flow.

3. Backlund Transformations.

The classical Darboux transformation, which leads to the well-known Bäcklund transformation for the KdV equation, also has a nice formulation in terms of transmutation operators, as follows. Consider the transformation of a potential u_1 to u_2 by the equation $(D^2+u_2)(D+v) = (D+v)(D^2+u_1)$.This operator identity reduces to the equation $(u_2-u_1 + 2v')D +(v'' +(u_2-u_1) - u_1') = 0$, hence to the pair

$$u_2-u_1 = -2v' \qquad\qquad v'' + v\,(u_2-u_1) - u_1' = 0. \qquad (3.1)$$

Substituting the first equation into the second we get $d/dx(v'-v^2-u_1) = 0$. Integrating this once we get $v' - v^2 - u_1$ =const. This is the Miura transformation that played a fundamental role in the early development of the KdV equation. Taking the constant to be ω^2 and introducing ψ by $v = d/dx \log 1/\psi$, we get a Schrödinger equation for ψ:

$$\psi'' + (u_1 + k_0^2)\psi = 0, \qquad\qquad v' - v^2 - u_1 = \omega^2. \qquad (3.2)$$

In other words, ψ is a solution of the Schrödinger equation with the original potential u_1. The new potential is then given by

$$u_2 - u_1 = -2 \frac{d}{dx} v = 2 \frac{d^2 \log \psi}{dx^2}$$

For example, if $u_1 = 0$ then ψ satisfies $\psi'' + \omega^2\psi = 0$. Taking $\psi = \cosh \omega x$ we get $u_2 = 2\omega^2 \mathrm{sech}\, \omega^2 x$. This is the well-known wave form of the solitary wave of the KdV equation

In order to get a time dependent solution of the KdV equation, we must also require D+v to satisfy the identity

$$\left(\frac{\partial}{\partial t} - B_2\right)(D+v) = (D+v)\left(\frac{\partial}{\partial t} - B_1\right) \tag{3.3}$$

where $B_j = B(u_j)$ is the Lax operator for the potential u_j: $B(u) = D^3 + 3/4(Du + uD)$. (For a derivation, see the next section.) By commuting all powers of D to the right in (3.3)(a rather tedious computation), we obtain an operator identity which is second order in D.)The coefficients of D^2 and D are consistent with the pair (3.1) and are satisfied automatically. Using (3.1) and the Miura transformation $v' - v^2 - u_1 = \omega^2$, the coefficient of D^0 simplifies to

$$v_t - \frac{1}{4} v_{xxx} + \frac{3}{2}(v^2 + \omega^2) v_x = 0 \tag{3.4}$$

This equation, modulo the term $\omega^2 v_x$, is known as the modified KdV equation. (The term ω^2 can be scaled away using a Gallilean transformation.)

The KdV equation is given by the commutation relation $[L, \partial/\partial t - B] = 0$. If u_1 satisfies the KdV equation and v satisfies the pair of equations (3.2, 3.4), then $u_2 = u_1 - 2 v_x$, is also a solution of the KdV equation.

4. The dressing method and the Kadomtsev-Petviashvili hierarchy.

The KdV equation is only one equation in an infinite hierarchy of commuting nonlinear flows, and furthermore is embedded in a class of two dimensional flows, known as the *Kadomtsev- Petviashvili* (KP) hierarchy. The entire KP hierarchy can be obtained by dressing the operators $\partial/\partial x_n - D^n$, n=2,3,...and $D = \partial/\partial x$ [20]. We introduce the hierarchy variable $\mathbf{x} = (x, x_2, x_3, \ldots)$. The kernels $K_\pm$ are formally functions of the hierarchy variable **x** and z. In order to make analytic sense of the expressions, one may set all but a finite number of the x_n equal to zero. Let $K_\pm$ and B_n be operators satisfying the relationship

$$\left(\frac{\partial}{\partial x_n} - B_n\right)(1+K_\pm) = (1+K_\pm)\left(\frac{\partial}{\partial x_n} - D^n\right)$$

The B_n are n^{th} order differential operators with leading term D^n. The process of obtaining the coefficients of the B_n is an algebraic one. The coefficients of the lower order derivatives are determined by applying the above operator identities to a function ψ and integrating all derivatives in the integrands by parts. When this is done one ends up with integral terms (non-local terms) and local (differential) operators on ψ coming from the boundary terms. The B_n are determined from the local operations; while the integral terms give differential equations for the kernels $K_\pm$. For example, the operators B_2 and B_3 are

$$B_2=D^2+u(x), \qquad\qquad B_3=D^3+(3/4)(uD+Du) + w,$$

where

$$u(x) = 2\frac{\partial K(x,x)}{\partial x} \qquad\qquad w(x) = \frac{3}{2}\left(\frac{\partial^2}{\partial x^2} - \frac{\partial^2}{\partial z^2} + u(x)\right) K(x,z)\Big|_{z=x}$$

The KP hierarchy is obtained from the commutation relations

$$\left[\frac{\partial}{\partial x_n} - B_n, \frac{\partial}{\partial x_m} - B_m\right] = 0$$

These follow immediately from the dressing relation and the obvious fact that these commutation relations hold for the bare operators $\partial/\partial x_n - D^n$. The KP equation itself arises from taking n=2 and m=3:

$$\left[\frac{\partial}{\partial y} - D^2 - u, \frac{\partial}{\partial t} - D^3 - \frac{3}{4}(uD+Du) - w\right] = 0.$$

In fact, working out this commutator, we get

$$\left[\frac{\partial}{\partial y} - D^2 - u, \frac{\partial}{\partial t} - D^3 - \frac{3}{4}(uD+Du) - w\right] =$$

$$\left(u_t - \frac{1}{4}(u_{xxx} + 6uu_x) - w_y + w_{xx} - \frac{3}{4}u_{xy}\right) + \left(2w_x - \frac{3}{2}u_y\right)D$$

Setting both terms equal to zero we get

$$u_t - \frac{1}{4}(u_{xxx} + 6uu_x) = w_y \quad \text{and } w_x = \frac{3}{4}u_y$$

The Kadomtsev-Petviashvili equation follows by differentiating this equation with respect to x:

$$(u_t - \frac{1}{4}(u_{xxx} + 6uu_x))_x = \frac{3}{4}u_{yy}$$

The KP equation was derived by this method by Zakharov and Shabat. Their procedure may be extended to derive all the equations in the KP hierarchy by working out the commutators for the operators B_n.

As in the case of the KdV equation, we get the commutation relations

$$[F, \frac{\partial}{\partial x_n} - D^n] = 0 \quad \text{for } n \geq 2$$

This leads to the infinite system of *linear* partial differential equations for the kernel f:

$$\frac{\partial f}{\partial x_n} - \frac{\partial^n f}{\partial x^n} + (-1)^n \frac{\partial^n f}{\partial z^n} = 0, \qquad n \geq 2$$

These equations determine the flow in the variables $x_2, x_3, \ldots$. Thus the factorization $(1+F)=(1+K_+)^{-1}(1+K_-)$ provides an alternative way of linearizing the flow. A special solution of this system of equations is given by $f(x,z) = e^{\xi(x,p) - \xi(z,q)}$ where (with $x_1=x$)

$$\xi(x,k) = \sum_{j=1}^{\infty} x_j k^j$$

With this choice of f one obtains the one soliton solution of the KP hierarchy. The parameters p and q are complex; note that f tends to zero exponentially as z tends to ∞ if $\mathrm{Re}\, p < 0 < \mathrm{Re}\, q$.

As in the case of the KdV equation, $\tau = \det(1+F_x)$. The potential u is again obtained from the formula $u=2(\log \tau)_{xx}$. Moreover all the coefficients of the differential operators B_n can be recovered from the derivatives of $\log \tau$ [7].

The wave function $e^{\xi(x,k)}$ satisfies the infinite set of equations $(\partial/\partial x_n - D^n)\, e^{\xi(x,k)} = 0$. The wave function for the KP hierarchy (also called the Baker Akhiezer function) is given by

$$w(\mathbf{x},k)=(1+K_+)e^{\xi(\mathbf{x},k)}= e^{\xi(\mathbf{x},k)} + \int_x^\infty K_+(\mathbf{x},z)e^{\xi(z,k)}dz \quad .$$

All of the known special multi-soliton solutions have been obtained, essentially, by evaluating the Fredholm determinant of the GLM equation. The Fredholm determinant method also gives a concrete representation of the τ function for the initial value problem for the KdV and KP I hierarchies. But it raises fundamental questions about the extent of the dressing method in the context of multi-dimensional isospectral problems. Inverse problems connected with multi-dimensional isospectral operators can lead to ∂-bar problems in which the wave function w(x,k) may not be analytic anywhere in the complex plane (cf. [3]). This happens in the case of the KP hierarchy for real values of the variables x_{2j} . This case has been called "KP II" by Ablowitz, Bar Yaacov, and Fokas, or the "stable" case by Zakharov and Shabat. The treatment of the general initial value problem for this case requires the use of the ∂-bar method. The dressing method, if it applies at all to this case, will have to be substantially modified. When the x_{2j} are imaginary (the "KP I", or "unstable," case), the wave function w is analytic in the left and right half k- planes, and the associated isospectral problem leads to a non-local Riemann-Hilbert problem. The solution of the initial value problem for KP I by the GLM equation has been discussed by Manakov [15].

The τ function for the multi-soliton solutions of KP I can be analytically continued to real x_{2j} to obtain multi-soliton solutions for KP II. In general, however, additional contraints must be placed on the parameters to ensure that the analytic continuation is real and positive. This is illustrated by some of the examples in [20].

Sato [21] and Date *et. al.* [7] developed a theory of the τ function for the KP hierarchy in terms of Grassman manifolds, and obtained the rational , multi-soliton and quasi-periodic solutions of Krichever. The rational and quasi-periodic solutions cannot be obtained directly from the inverse scattering theory, *viz*. from the Fredholm determinant of the scattering operator; Pöppe, however, has obtained the rational solutions as the determinant of a suitably chosen family of finite dimensional matrices [19].

References

1. Ablowitz, M. J. and Segur, H. *Solitons and the Inverse Scattering Transform* SIAM, Philadelphia, 1981

2. Ablowitz, M.J. and Fokas, A.S. "On the inverse scattering of the time dependent Schrödinger equation and the associated Kadomtsev-Petviashvili (I) equation," *Studies in Applied Mathematics,* **69** (1983), 211-228.

3. Ablowitz, M. J., Bar Yaacov, D., and Fokas, A.S. "On the inverse scattering transform for the Kadomtsev-Petviashvili Equation," *Studies in Applied Math.* **69** (1983), 135-143.

4. Agranovich, Z.S., and Marchenko, V.A. *The Inverse Problem of Scattering Theory* Gordon and Breach, New York, 1963.

5. Beals, R., Deift, P. and Tomei, C. *Direct and Inverse Scattering on the Line* Research Monographs in Mathematics, vol. 28, Amer. Math. Soc., Providence, 1988

6. Carroll, R. *Transmutation Theory and Applications,* North Holland, Amsterdam, 1985.

7. Date, Kashiwara, Jimbo, and Miwa, "Transformation groups for soliton equations, " *Proc. of RIMS symposium on nonlinear integrable systems*, ed. Jimbo and Miwa, World Scientific Publishing Co. Singapore, 1983.

8. Dodd, Eilbeck, Gibbon, and Morris, *Solitons and Nonlinear Equations*, Academic Press, New York, 1982.

9. Dyson, F.J. "Old and new approaches to the inverse scattering problem," *Studies in Math. Physics*, Princeton Series in Physics, eds. Lieb, Simon, Wightman, 1976; "Fredholm determinants and inverse scattering problems," *Comm. Math. Phys.* **47**, (1976) 171-183.

10. Gel'fand, I. M , Dikii,L "Fractional powers of operators and Hamiltonian systems," *Functional Analysis and its Applications*, **10**, (1976), 259-273.

11. Hirota, R. "Direct methods in soliton theory," in *Solitons*, Topics in Current Physics, 17; ed. R.K. Bullough and P.J. Caudrey, Springer-Verlag, Heidelberg, 1980.

12. Lax, P., and Phillips, R. *Scattering Theory* Academic Press, New York, 1967.

13. Lions, J.L."Quelques applications d'opérateurs de transmutation," *La Théorie des Équations aus Dérivées partielles,* C.N.R.S. Paris, 1956

14. Manakov, S.V. "The inverse scattering transform for time dependent Schrödinger equation and Kadomtsev-Petviashvili equation," *Physica* **3D**, (1981), 420-427.

15. Melin, A., "Operator methods for inverse scattering on the real line," *Comm. in Partial Diff. Equations,* **10** (1985), 677-766.

16. Novikov, Manakov, Pitaevski, and Zakharov, *Theory of Solitons*, Plenum Publishing, New York, 1984.

17. Oishi, S. "Relationship between Hirota's method and the inverse spectral method- the Korteweg- deVries equation's case" *Jour. Phys. Soc. Japan* **47** (1979), 1037-38.

18. Pöppe, C., "Construction of solutions of the sine-Gordon equation by means of Fredholm determinants," *Physica 9D* (1983), 103.; "The Fredholm Determinant method for the

KdV equations," *Physica* 13D (1984), 137-160.

19. _______"General determinants and the τ function for the Kadomtsev-Petviashvili hierarchy," to appear, *Inverse Problems*

20. Pöppe, Ch., and Sattinger, D.H. "Fredholm determinants and the τ function for the Kadomtsev-Petviashvili hierarchy," *Publ. RIMS*, Kyoto University, **24** (1988), 505-538.

21. Sato, M. "Soliton equations as dynamical systems on infinite dimensional Grassmann manifolds," *RIMS Kokyuroku*, **439** (1981) 30.

22. Zakharov, V. and Shabat, A.B.,"A scheme for integrating the nonlinear equations of mathematical physics by the method of the inverse scattering problem," *Functional Analysis and its Applications*, **8** (1974), 226-235.

5. Degenerate Parabolic Equations

Recent results on the Cauchy problem and initial traces for degenerate parabolic equations

E. DiBenedetto*
Northwestern University
Evanston, Illinois 60208 U.S.A.

1. Introduction.

We will present some recent results [14,40] on the Cauchy problem and existence of initial traces for the non-linear evolution equation

$$u_t - \operatorname{div}(|Du|^{p-2} Du) = 0, \qquad p > 1, \tag{1.1}$$

and indicate some open problems.

We will consider only non-negative solutions of (1.1) in the strip

$$S_T = \mathbf{R}^N \times (0,T); \quad N \geq 2; \qquad 0 < T < \infty.$$

The concept of solution will be made precise below, both for the cases $p \geq 2$ and $1 < p < 2$.

Here we state informally. the problems we study. These are best described having the heat equation as a guideline. In this context the theory has been developed by Thychonov [37], Widder [39], Aronson [3], Tacklind [36], Kamynin [23].

2. Heuristic results on the Cauchy problem.

The unique solution of the Cauchy problem for the heat equation

$$\begin{cases} u_t - \Delta u = 0 & \text{in } S_T \\ u(\cdot,0) = u_0(\cdot), & u_0 \geq 0 \end{cases} \tag{2.1}$$

is given by

$$u(x,t) = (4\pi t)^{-\frac{N}{2}} \int_{\mathbf{R}^N} e^{-\frac{|x-\xi|^2}{4t}} u_0(\xi) d\xi, \tag{2.2}$$

*Partially supported by NSF grant DMS-8802883.

as long as the integral converges. This will happen if, for example,

$$u_0 \in L^\infty_{\text{loc}}(\mathbf{R}^N) \quad \text{and} \quad u_0(x) \le Ce^{\frac{|x|^2}{4T}}, \tag{2.3}$$

for some constant C. This is a growth condition on $u_0(\cdot)$ as $|x| \to \infty$. The solution formula (2.2) implies that u_0 does not have to be a $L^\infty_{\text{loc}}(\mathbf{R}^N)$-function; it could be (modulo standard approximations) a function in $L^1_{\text{loc}}(\mathbf{R}^N)$ or even a Radon measure μ.

Precisely if μ satisfies

$$\int_{\mathbf{R}^N} e^{-a|x|^2} d\mu < \infty \tag{2.4}$$

for some $a > 0$, then the Cauchy problem (2.1) with μ_0 replaced by μ is uniquely solvable in the strip

$$S_T; \qquad T = \frac{1}{4a} \tag{2.5}$$

and the solution is given by

$$u(x,t) = (4\pi t)^{-\frac{N}{2}} \int_{\mathbf{R}^N} e^{-\frac{|x-\xi|^2}{4t}} d\mu; \qquad 0 < t < T. \tag{2.6}$$

Remark. From (2.4)–(2.5) it follows that the solution is *local in time*, and it is global in time if $a = 0$.

The same issue for (1.1) can be given the following heuristic answer (again we postpone the precise statement until later).

2-(i). The case $p > 2$. The Cauchy problem associated with (1.1) is solvable if the initial datum $x \to u_0(x)$ grows as $|x| \to \infty$ no faster than $|x|^{\frac{p}{p-2}}$, say for example if

$$u_0 \in L^\infty_{\text{loc}}(\mathbf{R}^N), \quad u_0(x) \le C(1 + c_0|x|^{\frac{p}{p-2}}) \qquad \text{for } |x| \gg 1, \tag{2.7}$$

for some constants C, c_0. In such a case the solution u is local in time and exists within the strip S_T, $T = D/c_0^{p-2}$ for a constant D depending only upon N, p. The solution is global in time if $c_0 = 0$. Also u_0 does not have to be in $L^\infty_{\text{loc}}(\mathbf{R}^N)$; it could be in $L^1_{\text{loc}}(\mathbf{R}^N)$ or even a Radon measure μ, with the growth condition (2.7) suitably rephrased (see §4) in terms of integral averages.

2-(ii). The case $1 < p < 2$. Let $u_0 \in L^1_{\text{loc}}(\mathbf{R}^N)$, $u_0 \ge 0$. The Cauchy problem associated with (1.1), $1 < p < 2$ and u_0 is always solvable, *globally in time*, regardless of how fast $x \to u_0(x)$ "grows" as $|x| \to \infty$.

The solutions are however less regular than those in the case $1 < p < 2$ and the concept of solution itself is rather delicate (see §5).

3. Heuristic results on initial traces.

Let u be a *given* non-negative classical solution of the heat equation in $S_{(T+\varepsilon)}$, for some $0 < T < \infty$ and some $\varepsilon \in (0,1)$.

We whish to reconstruct the unknown "initial datum" $u(.,0)$ from the knowledge of u. For this let $\tau \in \left(0, \frac{T}{2}\right)$, fix any $\rho > 0$ and consider the ball $B_\rho \equiv \{|x| < \rho\}$. Let $(x,t) \to \bar{u}(x,t)$ be the unique solution of the Cauchy problem

$$\bar{u}_t - \Delta\bar{u} = 0 \qquad \text{in } \mathbf{R}^N \times (\tau, \infty)$$
$$\bar{u}(x,\tau) = \begin{cases} u(x,\tau) & x \in B_\rho \\ 0 & |x| \geq \rho, \end{cases}$$

given by

$$\bar{u}(x,t) = [4\pi(t-\tau)]^{-\frac{N}{2}} \int_{B_\rho} e^{-\frac{|x-\xi|}{4(t-\tau)}} u(\xi,\tau)d\xi. \tag{3.1}$$

Since $\bar{u} \leq u$ in $\mathbf{R}^N \times [\tau, T]$, by the maximum principle, calculating (3.1) for $x = 0$ and $t = T$ yields

$$\int_{B_\rho} u(x,\tau)dx \leq (4\pi T)^{\frac{N}{2}} e^{\frac{\rho^2}{4(T-\tau)}} u(0,T). \tag{3.2}$$

Inequality (3.2) is referred to as a Harnack-type inequality for solutions of the heat equation in S_T. It says that the net $\{u_\tau(x)\} \equiv \{u(x,\tau)\}$ is equibounded in $L^1_{\text{loc}}(\mathbf{R}^N)$ and (passing to a subnet) as $\tau \to 0$ it converges in the sense of measures to a Radon measure μ that "grows" as $|x| \to \infty$ no faster then $e^{\frac{|x|^2}{4T}}$, or more precisely satisfying (2.4).

For the non-linear equation (1.1) the question of initial trace can be outlined as follows.

3-(i). The case $p > 2$. Let u be a non-negative solution of (1.1) in S_T.The "initial datum" for u is a Radon measure μ "growing," as $|x| \to \infty$ no faster then $|x|^{\frac{p}{p-2}}$. Precisely $\forall\rho > 1$,

$$\frac{1}{\rho^N} \int_{B_\rho} d\mu \leq \text{const} \left[1 + \left(\frac{\rho^p}{T}\right)^{\frac{1}{p-2}}\right]$$

where const depends upon T, and the value of the solution u at $(0,T)$.

3-(ii). The case $1 < p < 2$. Also in this case there exist a Radon measure μ as initial trace. The solutions here are taken with no growth restriction as $|x| \to \infty$. The corresponding initial traces exhibit no particular growth restriction as $|x| \to \infty$.

4. Some rigorous results; $p > 2$.

A non-negative measurable function $u : S_T \to \mathbf{R}$ is a weak solution of (1.1) in S_T if for every bounded open set $\Omega \subset \mathbf{R}^N$

$$u \in C(0,T;L^1(\Omega)) \cap L^p(0,T;W^{1,p}(\Omega)) \tag{4.1}$$

$$\int_\Omega u(x,\tau)\varphi(x,\tau)dx\Big|_{t_1}^{t_2} + \int_{t_1}^{t_2}\int_\Omega \{-u\varphi_t + |Du|^{p-2}DuD\varphi\}dxd\tau = 0 \tag{4.2}$$

for all $0 < t_1 < t_2 \leq T$ and all testing functions

$$\varphi \in W^{1,\infty}(0,T;L^\infty(\Omega)) \cap L^\infty(0,T;W_0^{1,\infty}(\Omega)). \tag{4.3}$$

Analogously, if μ is a Radon measure, a weak solution of the Cauchy problem

$$\begin{cases} u_t - \operatorname{div}|Du|^{p-2}Du = 0 & \text{in } S_T \\ u(\cdot,0) = \mu \end{cases} \tag{4.4}$$

is defined as in (4.1)–(4.3) where t_1 is allowed to be zero and

$$\int_\Omega u(x,\tau)\varphi(x,\tau)dx\Big|_{\tau=0} \equiv \int_\Omega \varphi(x,0)d\mu. \tag{4.5}$$

If $\mu \equiv u_0 \in C_0^\infty(\mathbf{R}^N)$, the corresponding solution of (4.3) can be constructed by solving first the boundary value problem

$$\begin{cases} \frac{\partial}{\partial t}u_n - \operatorname{div}(|Du_n|^{p-2}Du_n) = 0 & \text{in } Q_n \equiv \{|x| < n\} \times \{0,\infty\}, n \in \mathbf{N} \\ u_n(x,0) = u_0(x), & |x| < n \\ u_n(x,t) = 0, & |x| = n; \quad t > 0. \end{cases} \tag{4.6$_n$}$$

These, in turn, can be solved by Galerkin-type procedure (see [26] for details). Standard energy estimates give

$$\|u_{n,t}\|_{2,Q_n} + \|Du_n\|_{p,Q_n} \leq \|Du_0\|_{p,\mathbf{R}^N} + \|u_0\|_{2,\mathbf{R}^N} \tag{4.7}$$

uniformly in n.

Moreover, by the results of [15] and [16],

$$(x,t) \to Du_n(x,t) \in C^\alpha(\mathbf{R}^N \times (\tau,\infty)) \qquad \text{uniformly in } n, \tag{4.8}$$

for some $\alpha \in (0,1)$ depending upon $\|u_0\|_{2,\mathbf{R}^N} + \|Du_0\|_{p,\mathbf{R}^N}$. Here we view at u_n as defined in the whole strip S_T by extending them to be zero for $|x| \geq n$.

These estimates supply the necessary compactness to pass to the limit as $n \to \infty$ in the weak formulation of (4.6$_n$) to obtain solutions of (4.2).

Next we allow for initial data $u_0 \in L^1_{\rm loc}(\mathbf{R}^N)$ growing as $|x| \to \infty$. Such a "growth" is measured by

$$\left|\left\|u_0\right\|\right|_r = \sup_{\rho \geq r} \fint_{B_\rho} \frac{u_0(x)}{\rho^{\frac{p}{p-2}}}\, dx, \qquad \text{for some } r > 0. \tag{4.9}$$

Here, and in what follows, for a σ-finite non-negative Borel measure μ in $\mathbf{R}^N$

$$\fint_{B_\rho} d\mu = \rho^{-N} \int_{B_\rho} d\mu.$$

Let us assume first that $u_0 \in C^\infty(\mathbf{R}^N)$ and

$$\left|\left\|u_0\right\|\right|_r < \infty \qquad \text{for some } r > 0. \tag{4.10}$$

Condition (4.10) says that $x \to u_0(x)$ grows as $|x| \to \infty$ no faster than $|x|^{\frac{p}{p-2}}$.

Let $\{u_{0,j}\}$ be a sequence of functions in $C_0^\infty(\mathbf{R}^N)$ to u_0 convergent in $L^1_{loc}(\mathbf{R}^N)$ to u_o as $j \to \infty$, and let u_j be the corresponding solutions of (4.4) with μ replaced by $u_{0,j}$. Their weak formulation reads

$$\begin{aligned} \int_{\mathbf{R}^N} u_j(t)\varphi(t)dx + \int_0^t \int_{\mathbf{R}^N} \{-u_j\varphi_t + |Du_j|^{p-2} Du_j D\varphi\} dx d\tau \\ = \int_{\mathbf{R}^N} u_{0,j}\varphi(0)dx; \qquad \forall t > 0, \end{aligned} \tag{4.11}$$

$$\forall \varphi \in C^1([0,\infty; C_0^\infty(\mathbf{R}^N)). \tag{4.12}$$

The necessary compactness to let $j \to \infty$ in (4.8) consist of

(a) A *local* sup bound for $\{u_j\}$ uniform in j
(b) A local Hölder estimate of $\{Du_j\}$ uniform in j
(c) An estimate "up to zero" of Du_j of the type $\forall t > 0$; $\forall \rho > 0$

$$\int_0^t \int_{B_\rho} |Du_j|^{p-1} dx d\tau \leq \text{const}(\rho, t).$$

Estimates of this kind are possible only *locally in time* and are supplied by the following theorem:

THEOREM 1. *There exist constants C_i, $i = 0,1,2,3,4$ depending only upon N and p such that setting*

$$T_r(u_0) = C_0 \left|\left\|u_0\right\|\right|_r^{-(p-2)}$$

there holds, $\forall 0 < t < T_r(u_0)$, $\forall r > 0$, $\forall \rho \geq r$, $\forall j \in \mathbf{N}$,

$$\text{(I)} \qquad \left|\left\|u_j(\cdot,t)\right\|\right|_r \leq C_j \left|\left\|u_0\right\|\right|_r,$$

$$\text{(II)} \qquad \|u_j(\cdot,t)\|_{\infty,B_\rho} \leq C_2 t^{-\frac{N}{\kappa}} \rho^{\frac{p}{p-2}} \left|\left\|u_0\right\|\right|_r^{p/\kappa}; \qquad \kappa = N(p-2)+p$$

$$\text{(III)} \qquad \|Du_j(\cdot,t)\|_{\infty,B_\rho} \leq C_3 t^{-\frac{(N+1)}{\kappa}} \rho^{\frac{2}{p-2}} \left|\left\|u_0\right\|\right|_r^{\frac{2}{\kappa}}$$

$$\text{(IV)} \qquad \int_0^t \int_{B_\rho} |Du_j|^{p-1} dx d\tau \leq C_4 t^{\frac{1}{\kappa}} \rho^{1+\frac{\kappa}{p-2}} \left|\left\|u_0\right\|\right|_r^{1+\frac{p-2}{\kappa}}$$

Given the sup-estimate (III),from [15], C^α estimates on Du_j follow .

The theorem yields existence of a solution in the strip $S_{T_r}(u_0)$. Thus the solutions are *only local* in time. Since the number $r > 0$ is arbitrary, by letting $r \to \infty$ in the definition of $T_r(u_0)$ we obtain the largest strip $S_{T_\infty(u_0)}$ within which a solution can be found. In particular if

$$(4.13) \qquad \left|\left\|u_0\right\|\right|_r \longrightarrow 0 \quad \text{as } r \to \infty$$

then $T_r(u_0) \to +\infty$ and the solution is global in time.

Condition (4.13) occurs, loosely speaking, if u_0 grows slower than $|x|^{\frac{p}{p-2}}$ as $|x| \to \infty$.

Given the nature of the estimates in Theorem 1, u_0 could be replaced by a function in $L^1_{\text{loc}}(\mathbf{R}^N)$ or even a Radon measure μ satisfying

$$(4.14) \qquad \left|\left\|\mu\right\|\right| = \sup_{\rho \geq r} \fint_{B_\rho} \frac{d\mu}{\rho^{\frac{p}{p-2}}} < \infty \qquad \text{for some } r > 0.$$

If $u_0 \in L^1_{\text{loc}}(\mathbf{R}^N)$ and if $\left|\left\|u_0\right\|\right|_r < \infty$, then the constructed solutions are unique (see [14]). It is an open problem to establish uniqueness when the initial datum is a Radon measure μ satisfying (4.14).

We turn to the question of initial traces. Motivated by the Harnack type inequality (3.2) for the heat equation, we seek a similar estimate for weak solutions of (1.1), $p > 2$.

We let u be a non-negative weak solution of (1.1) in S_T, $0 < T < \infty$, with no reference to initial data (see (4.1)–(4.3)). The two basic facts that permit to establish existence of initial traces are

A) Harnack type inequality

There exists a constant C depending only upon N and p such that

$$\forall \tau \in \left(0, \frac{T}{2}\right), \qquad \forall \rho > 0$$

$$(4.15) \qquad \fint_{B_\rho} u(x,\tau)dx \leq C\left\{\left(\frac{\rho^p}{T}\right)^{\frac{1}{p-2}} + \left(\frac{T}{\rho^p}\right)^{\frac{N}{p}} [u(0,T)]^{\frac{\kappa}{p}}\right\}$$

$$\kappa = N(p-2)+p.$$

B) Spreading of positivity

$$\forall \varepsilon \in (0,1), \qquad \forall \rho > 1, \qquad \forall 0 < \tau < t < T,$$

$$\fint_{B_{(1+\varepsilon)\rho}} u(x,t)dx \geq \fint_{B_\rho} u(x,\tau) \left\{ (1+\varepsilon)^{-N} - \frac{C}{\varepsilon}\left(\frac{t}{\rho^p}\right)^{\frac{1}{\kappa}} \right\} \tag{4.16}$$

where C depends upon u and ρ but it is independent of t.

Inequality (4.12) is used exactly as in the case of the heat equation. If $\rho > 0$ is fixed, from the $L^1_{\text{loc}}(\mathbf{R}^N)$-bounded net $\{u(\tau)\} \equiv \{u(\cdot,\tau)\}$ we may extract a subnet indexed with τ' such that

$$u(\tau') \longrightarrow \mu \qquad \text{in the sense of measures as } \tau' \to 0.$$

Suppose now that for another subnet indexed with τ''

$$u(\tau'') \longrightarrow \nu \qquad \text{in the sense of measures as } \tau'' \to 0,$$

and assume that $\mu \neq \nu$. From (4.16) for ρ fixed let $t \to 0$ along τ' to get

$$\fint_{B_{(1+\varepsilon)\rho}} d\nu \geq (1+\varepsilon)^{-N} \fint_{B_\rho} u(x,\tau)dx.$$

Now let $\varepsilon \to 0$ and then $\tau \to 0$ along τ'. This gives

$$\int_{B_\rho} d\mu \leq \int_{B_\rho} d\nu; \qquad \forall \rho > 0.$$

Hence $\mu \equiv \nu$, and for the entire net $\{u(\tau)\}$ we have as $\tau \to 0$

$$u(\tau) \longrightarrow \mu \qquad \text{in the sense of measures.}$$

Let now $\tau \to 0$ in (4.15) to find

$$\fint_{B_\rho} d\mu \leq C\left\{ \left(\frac{\rho^p}{T}\right)^{\frac{1}{p-2}} + \left(\frac{T}{\rho^p}\right)^{\frac{N}{p}} [u(0,T)]^{\frac{\kappa}{p}} \right\}.$$

This says that the initial trace μ of u "grows" as $|x| \to \infty$ no faster than $|x|^{\frac{p}{p-2}}$.

The results are sharp in view of some available explicit solutions of (1.1) (see [14]).

5. Some rigorous results, $1 < p < 2$.

Consider the Cauchy problem

$$u_t - \operatorname{div} |Du|^{p-2}Du = 0 \qquad \text{in } \mathbf{R}^N \times (0,T), \quad 0 < T < \infty; \quad 1 < p < 2, \tag{5.1}$$

$$u(\cdot,0) = u_0(\cdot) \in L^1_{\text{loc}}(\mathbf{R}^N); \qquad u_0 \geq 0. \tag{5.2}$$

We discuss two subcases, according to some possible relationship between the "regularity of u_0" and the "size of p."

5-(i). $u_0 \in L^r_{\mathrm{loc}}(\mathbf{R}^N)$; $p > \max\left\{1; \frac{2N}{N+2}\right\}$.

In this case u_0 could be merely in $L^1_{\mathrm{loc}}(\mathbf{R}^N)$ if p is sufficiently large, or p could be arbitrarily close to 1 provided $u_0 \in L^r_{\mathrm{loc}}(\mathbf{R}^N)$ and r is sufficiently large. The theory here develops along the lines of Section 4. There are major differences, however, and we will point them out as we come to them.

The definition of weak solution of (5.1) in S_T, $\forall 0 < T < \infty$ and the corresponding concept of solution of the Cauchy problem (5.1)–(5.2) here are exactly the same as in (4.1)–(4.5) with the obvious modifications.

Solutions to the Cauchy problem can be constructed by starting with

$$u_0 \in C_0^\infty(\mathbf{R}^N) \tag{5.3}$$

and solving a sequence of "chopped" problems in all analogous to $(4.6)_n$. Estimates (4.7) continue to hold uniformly in n since $u_0 \in C_0^\infty(\mathbf{R}^N)$.

Since $u_0 \in C_0^\infty(\mathbf{R}^N)$, we have the obvious energy and maximum principle estimates, uniform in n

$$\|u_n(\cdot,t)\|_{\infty,\mathbf{R}^N} \le \|(u_0)\|_{\infty,\mathbf{R}^N}; \quad \|Du_n\|^p_{p,S_T} \le \|u_0\|^2_{2,\mathbf{R}^N}; \quad \forall 0 < T < \infty, \tag{5.5}$$

$$\left\|\frac{\partial}{\partial t}u_n\right\|^2_{2,S_T} \le \|Du_0\|^p_{p,\mathbf{R}^N}; \qquad \forall 0 < T < \infty.$$

Therefore by possibly passing to a subsequence, $\forall 0 < T < \infty$

$$\begin{aligned} u_n &\longrightarrow u \qquad \text{strongly in } L^2_{\mathrm{loc}}(S_T)\\ |Du_n|^{p-2}Du_n &\longrightarrow \overrightarrow{\chi} \in L_{\mathrm{loc}}^{\frac{p}{p-1}}(S_T) \qquad \text{weakly in } L^p_{\mathrm{loc}}(S_T)\\ u &\in L^p(0,T;W^{1,p}_{\mathrm{loc}}(\mathbf{R}^N)), \qquad u_t \in L^2_{\mathrm{loc}}(S_T). \end{aligned}$$

To let $n \to \infty$ pick φ as in (4.3) and write $(4.6)_n$ (for $1 < p < 2$) in the weak form

$$\int_\Omega u_n(x,\tau)\varphi(x,\tau)dx\Big|_0^t + \int_0^t\int_\Omega \{-u_n\varphi_t + |Du_n|^{p-2}Du_nD\varphi\}dxd\tau = 0 \tag{5.5}$$

$\forall 0 < t < \infty$; $\quad \forall n \ge n_0, n_o$ so large that $\Omega \subset \{|x| < n_0\}$.

Letting $n \to \infty$ in (5.6) we find

$$\int_{B_\rho} u(x,t)\varphi(x,t)dx + \int_0^t\int_{B_\rho}\{-u\varphi_t + \overrightarrow{\chi}D\varphi\}dxd\tau = \int_{B_\rho} u_0\varphi(x,0)dx, \tag{5.7}$$

$\forall \rho > 0$, $\forall \rho$ as in (4.3), $\Omega \subset B_\rho$.

Fix $\rho > 0$ and $\forall n > \rho$ in (5.6) choose $\varphi = u_n\zeta$, where $x \to \zeta(x)$ is a non-negative piecewise smooth function vanishing for $|x| \geq \rho$. Standard calculations give

$$\lim_{n\to\infty} \int_0^t \int_{B_\rho} |Du_n|^p \zeta dx d\tau = \frac{1}{2} \int_{B_\rho} (u_0^2 - u^2(t))\zeta dx - \int_0^t \int_{B_\rho} (\overrightarrow{\chi} D\zeta) u dx d\tau.$$

On the other hand, by taking $\rho = u\zeta$ in (5.7) we obtain

$$\frac{1}{2} \int_{B_\rho} (u_0^2 - u^2(t))\zeta dx - \int_0^t \int_{B_\rho} (\overrightarrow{\chi} D\zeta) u dx d\tau = \int_0^t \int_{B_\rho} (\overrightarrow{\chi} Du)\zeta dx d\tau.$$

Therefore

$$\lim_{n\to\infty} \int_0^t \int_{B_\rho} |Du_n|^p \zeta dx d\tau = \int_0^t \int_{B_\rho} (\overrightarrow{\chi} \cdot Du)\zeta dx d\tau. \tag{5.8}$$

If $\zeta \geq 0$, $\forall n \geq n_0$, $\forall \rho < n$, $\forall \rho \in C_0^\infty(S_T)$,

$$\int_0^t \int_{B_\rho} (|Du_n|^{p-2} Du_n - |D\varphi|^{p-2} D\varphi)(Du_n - D\varphi)\zeta dx d\tau \geq 0.$$

Expanding this expression, letting $n \to \infty$ and using (5.8) gives

$$\int_0^t \int_{B_\rho} (\overrightarrow{\chi} - |D\varphi|^{p-2} D\varphi)(Du - D\varphi)\zeta dx d\tau \geq 0$$

$\forall \varphi \in C_0^\infty(S_T)$. Hence $\overrightarrow{\chi} \equiv |Du|^{p-2} Du$ by Minty's lemma [29].

We conclude that if $u_0 \in C_0^\infty(\mathbf{R}^N)$, the Cauchy problem (5.1)–(5.2) has a solution n in the sense of the integral identity (5.7) satisfying estimates (5.5).

Next we give an initial datum

$$u_0 \in L^r_{\text{loc}}(\mathbf{R}^N), \qquad r \geq 1 \tag{5.9}$$

and assume that

$$p > \max\left\{1; \frac{2N}{N+r}\right\}. \tag{5.10}$$

Let $\{u_{0,j}\}$ be a sequence of functions in $C_0^\infty(\mathbf{R}^N)$ such that

$$u_{0,j} \longrightarrow u_0 \qquad \text{in } L^r_{\text{loc}}(\mathbf{R}^N), \quad r \geq 1,$$

and let $(x,t) \to u_j(x,t)$ be the solutions of the Cauchy problem (5.1)–(5.2) with initial datum $u_{0,j}$. The sequence $\{u_j\}$ satisfies

$$(5.11)\qquad \begin{cases} \frac{\partial}{\partial t} u_j \in L^2_{\mathrm{loc}}(S_T), & \forall 0 < T < \infty \\ |Du_j|^p \in L^p(0,T; L^p_{\mathrm{loc}}(\mathbf{R}^N)) & \\ u_j \in L^\infty(S_T), & \end{cases}$$

$$(5.12)\qquad \int_\Omega u_j(x,\tau)\varphi(x,\tau)dx\Big|_0^t + \int_0^t\int_\Omega \{-u_j\varphi_t + |Du_j|^{p-2}Du_jD\varphi\}dxd\tau = 0$$

$\forall$ bounded open sets $\Omega \subset \mathbf{R}^N$; $\forall 0 < t < \infty$; $\forall \varphi$ as in (4.3)

$u_j(x,0) = u_{0,j}(x) \quad \forall x \in \mathbf{R}^N$.

The quantitative estimates corresponding to (5.11) are *dependent upon j*.

The necessary compactness to let $j \to \infty$ in (5.12) consists of

(a) A local Hölder estimate on $\{u_j\}$ uniform in j

(b) An estimate of $|Du_j|$ in $L^p_{\mathrm{loc}}(S_T)$ uniform in j

(c) Estimates on $\{u_j\}$ and $\{Du_j\}$ "up to zero" of the type

$$\int_0^t\int_{B_\rho} |Du_j|^{p-1}dxd\tau \le \mathrm{const}(\rho,t,u_0);$$

$$\int_0^t\int_{B_\rho} |u_j|dxd\tau \le \mathrm{const}(\rho,t,u_0).$$

These are given by the following

THEOREM 2. *There exist constants C_i, $i = 1$ depending only upon N and p and independent of j such that*

$$(5.13)\qquad \forall 0 < t < \infty; \qquad \forall p > 0; \qquad \forall j \in \mathbf{N}$$

$$\|u_j(\cdot,t)\|_{\infty,B_\rho} \le C_1 \left\{ t^{-\frac{N}{\kappa_r}} \left(\int_{B_{2\rho}} u_0^r dx \right)^{\frac{p}{\kappa_r}} + \left(\frac{t}{\rho^p}\right)^{\frac{1}{2-p}} \right\}$$

$$(5.14)\qquad \kappa_r = N(p-2) + rp$$

$$(5.15)\qquad \sup_{0<\tau<t} \int_{B_\rho} u_j(x,\tau)dx \le C_2 \left\{ \int_{B_{2\rho}} dx + \left(\frac{t}{\rho^\kappa}\right)^{\frac{1}{2-p}} \right\},$$

(5.16) $$\kappa = \kappa_1 \equiv N(p-2) + p.$$

$$\int_0^t \int_{B_\rho} |Du_j|^{p-1} dx d\tau \le C_3 \left(\frac{t}{\rho^\kappa}\right)^{\frac{1}{p}} \cdot \left\{ \int_{B_{2\rho}} u_0 dx + \left(\frac{t}{\rho^\kappa}\right)^{\frac{1}{2-p}} \right\}^{\frac{2}{p}(p-1)}.$$

From (5.13) and [16] it follows that

$$u_j \in C^\alpha_{\text{loc}}(S_T) \qquad \forall 0 < T < \infty \text{ uniformly in } j, \text{ for some } \alpha \in (0,1).$$

This combined with (5.15) implies that, for a subsequence indexed again with j,

$$u_j \to u \text{ strongly in } L^1(0,T; L^1_{\text{loc}}(\mathbf{R}^N))$$
$$u_j \to u \textit{ uniformly} \text{ on compact subset of } S_T.$$

In particular writing (5.12) against testing functions

$$\varphi(x,\tau) = (\tau - \sigma)_+ u_j (\rho^2 - |x|^2)_+ \qquad \sigma \in (0,1), \rho > 0$$

we see that

$$|Du_j| \in L^p_{\text{loc}}(S_T)$$

so that $|Du| \in L^p_{\text{loc}}(S_T)$ by lower semi-continuity. Finally, for a subsequence indexed again with j

$$|Du_j|^{p-2} Du_j \rightharpoonup \overrightarrow{\chi}; \qquad \text{weakly in } L^{\frac{p}{p-1}}_{\text{loc}}(S_T).$$

Letting $j \to \infty$ in (5.12) gives

(5.18) $$\int_\Omega u(x,t)\varphi(x,t)dx + \int_0^t \int_\Omega \{-u\varphi_t + \overrightarrow{\chi} D\varphi\} dx d\tau = \int_\Omega u_0(x)\varphi(x,0)dx,$$

$\forall \varphi$ as in (4.3).

The identification $\chi \equiv |Du|^{p-2}Du$ here is carried within the strip $\mathbf{R}^N \times (\sigma, T)$, $\forall \sigma \in (0,T)$. We first write (5.12) and (5.18) in the form (4.2) with $t_1 = \sigma$, $t_2 = T$ and then proceed as before, modulo taking instead of u as a testing function, its Steklov time averages

$$u_h(x,1) = \frac{1}{h} \int_t^{t+h} u(x,\tau) d\tau,$$

by a standard approximation process.

The key fact in this construction process is estimate (5.13), which in turn implies (by the results of [16]) that u_j are locally Hölder continuous in S_T.

We notice that all the estimates of Theorem 2 are of *local* nature and no restriction on the growth of $x \to u_0(x)$ is required when $|x| \to \infty$. In turn the solutions constructed exhibit no particular growth as $|x| \to \infty$.

Estimate (5.13) is sharp; indeed in [40] a counterexample is given that shows that if $r = 1$ and $p = \frac{2N}{N+1}$, then $u_0 \in L^1(\mathbf{R}^N)$ (indeed, u_0 compactly supported) does not imply that $x \to u(x,t) \in L^\infty_{\text{loc}}(\mathbf{R}^N)$.

5-(ii). $u_0 \in L^1_{\text{loc}}(\mathbf{R}^N)$, $p \leq \frac{2N}{N+1}$..

As remarked previously, one does not have here a sup-estimate and consequently an estimate of the type

$$|Du| \in L^p_{\text{loc}}(S_T)$$

cannot be inferred. One still would have that (formally)

$$|Du| \in L^{p-1}_{\text{loc}}(S_T);$$

however, since $0 < p-1 < 1$, L^{p-1} is not a Banach space and it is not clear that the symbol Du means an operation of "weak derivative."

We introduce a new notion of weak solution motivated by the fact that heuristically $|Du| \notin L^p_{\text{loc}}(S_T)$ only because u fails to be locally bounded.

Define

$$\text{(5.19)} \qquad X_{\text{loc}}(S_T) \equiv L^p_{\text{loc}}(0,T; W^{1,p}_{\text{loc}}(\mathbf{R}^N) cap L^\infty_{\text{loc}}(S_T)$$

$$\overset{\circ}{X}_{\text{loc}}(S_T) \equiv \{\varphi \in X_{\text{loc}}(S_T) \mid \exists r > 0 : \varphi \in L^p_{\text{loc}}(0,T; \overset{\circ}{W}{}^{1,p}(|x| < r)\}.$$

For $f \in L^1_{\text{loc}}(S_T)$, $\forall k > 0$ set

$$f_k = \begin{cases} f & \text{if } f < k \\ k & \text{if } f \geq k. \end{cases}$$

A measurable function $u : S_T \to \mathbf{R}^+$ is a *local* weak solution of

$$\text{(5.20)} \qquad u_t - \operatorname{div} |Du|^{p-2} Du = 0 \qquad \text{in } S_T, \quad 1 < p < 2$$

if

$$\text{(5.21)} \qquad \begin{cases} u \in C(0,T; L^1_{\text{loc}}(\mathbf{R}^N)) \\ |Du_k| \in L^p_{\text{loc}}(S_T), & \forall k > 0 \\ \frac{\partial}{\partial t} u_k \in L^1_{\text{loc}}(S_T), & \forall k > 0, \end{cases}$$

$$\text{(5.22)} \qquad \forall \varphi \in \overset{\circ}{X}_{\text{loc}}(S_T), \qquad \forall 0 < s < t \leq T,$$

$$\int_s^t \int_{\mathbf{R}^N} \{u_t(\varphi - u)_+ + |Du|^{p-2} Du D(\varphi - u)_+\} dx d\tau = 0.$$

This notion eliminates, roughly speaking, those sets where "u is large" and might fail to be regular.

If $u_t \in L^1_{\text{loc}}(S_T)$ and $|Du| \in L^p_{\text{loc}}(S_T)$, then the notion coincides with the standard notion of distributional solutions. We prove furthermore that the truncations u_k are distributional supersolutions of (5.20) (see [40]).

Starting with an initial datum $u_0 \in L^1_{\text{loc}}(\mathbf{R}^N)$, $\quad u_0 \geq 0$, we construct solutions of (5.20) in the sense specified above which in addition satisfy

$$\overline{\lim_{k\to\infty}} \iint_{K\cap[k<u<\gamma k]} |Du|^p \frac{1}{u}\, dx d\tau = 0 \tag{5.23}$$

for every compact subset K of S_T, $\forall k > 0$ and $\forall \gamma > 1$.

As observed before we have no sup-estimates nor estimates of $|Du|$ as an element of $L^p_{\text{loc}}(S_T)$; for this the construction relies on rather delicate weak compactness arguments for which we refer to [40].

An interesting fact is that the initial datum u_0 is taken in the sense of $L^1_{\text{loc}}(\mathbf{R}^N)$, i.e.,

$$u(\cdot,t) \longrightarrow u_0(\cdot) \qquad \text{in } L^1_{\text{loc}}(\mathbf{R}^N) \text{ as } t \searrow 0.$$

We also prove that solutions in the sense of (5.21)–(5.22) are unique. Precisely

THEOREM 3. *Let u_1, u_2 be two weak solutions of (5.20) in S_T in the sense of (5.21)–(5.22) and satisfying (5.23). Then if*

$$(u_1 - u_2)(\cdot,t) \longrightarrow 0 \qquad \text{in } L^1_{\text{loc}}(\mathbf{R}^N) \text{ as } t \to 0,$$

then $u_1 \equiv u_2$ in S_T.

We finally come to the question of initial traces and their uniqueness. The theory here relies on the following Harnack-type estimates valid for any non-negative solution of (5.20) in S_T in the sense of (5.21)–(5.22); no reference to initial data is made.

HARNACK-TYPE ESTIMATE 1

$$\exists \gamma = \gamma(N,p) \text{ such that } \forall \rho > 0, \forall 0 < s < t \leq T, \tag{5.24}$$

$$\sup_{\tau\in(s,t)} \int_{B_\rho} u(x,\tau)dx \leq \gamma \left\{ \int_{B_{2\rho}} u(x,t)dx + \left(\frac{t-s}{\rho^\kappa}\right)^{\frac{1}{2-p}} \right\}$$

$$\kappa = N(p-2) + p.$$

HARNACK-TYPE ESTIMATE 2

$$\forall 0 < s < t \leq T; \quad \forall \sigma \in (0,1); \quad \forall \delta \in (0,1) \tag{5.25}$$

$$\int_{B_\rho} u(x,s)dx \leq \int_{B_{(1+\sigma)\rho}} u(x,t)dx + \delta \int_{B_{2(1+\sigma)\rho}} u(x,t)dx + \frac{\gamma(N,p<\delta)}{\sigma^{\frac{p}{2-p}}} \left(\frac{t}{\rho^\kappa}\right)^{\frac{1}{2-p}}.$$

Estimate (5.24) is used to establish existence of an initial trace and estimate (5.25) to prove their uniqueness along the lines of an argument similar to that in the case $p > 2$. We remark that (5.24) could be deduced from (5.25) by interpolation.

References

[1] N.D. Alikakos, R. Rostamian, Gradient estimates for degenerate diffusion equations II, Proc. Royal Soc. Edinburgh, Sect. A, 91, (1981/82), 335–346.

[2] S.N. Antonsev, Axially symmetric problems of gas dynamics with free boundaries, Doklady Akad. Nauk SSSR Tom. 216 (1974) #3, pp. 473–476.

[3] D.G. Aronson, Widder's inversion theorem and the initial distribution problem, SIAM J. Math. Anal. vol.12 #4 (1981), pp. 639–651.

[4] D.G. Aronson-L.A. Caffarelli, The initial trace of a solution of the porous medium equation, Trans. AMS 280 (1983) #1, pp. 351–366.

[5] P. Baras, M. Pierre, Singularites eliminables pour des equations semi-lineaires, Ann. Inst. Fourier (Grenoble) #34, (1884), pp.185–206.

[6] P. Baras, M. Pierre, Problemes paraboliques semi-lineaires avec donnees measures, Applicable Anal. #18 (1984), pp. 111–149.

[7] G.I. Barenblatt, On some unsteady motions of a liquid or a gas in a porous medium, Prikl. Mat. Mekh. 16 (1952), pp. 67–78.

[8] P. Benilan et al., personal communication .

[9] P. Benilan, M.G. Crandall, Regularizing effects of homogeneous evolution equations, MRC Tech. Rep. # 2076, Madison, WI (1980).

[10] P. Benilan-M.G. Crandall-M. Pierre, Solutions of the porous medium equation in R^N under optimal conditions on initial values, Indiana Univ. Math. Jour. #33 (1984), pp. 51–87.

[11] L. Boccardo, T. Gallouët, Non linear elliptic and parabolic equations involving measure data, Funct.Anal. (to appear).

[12] B.E.J. Dahlberg-C.E. Kenig, Non-negative solutions of generalized porous medium equations, Revista Matematica Iberoamericana Vol.2 #3 (1986), pp. 267–305.

[13] B.E.J. Dahlberg-C.E. Kenig, Non-negative solutions of the porous medium equation, Comm. Part. Diff. Equ. #9 (1984), pp. 409–437.

[14] E. DiBenedetto, M.A. Herrero, On the Cauchy problem and initial traces for a degenerate parabolic equation, to appear, Trans. Amer. Math. Soc.

[15] E. DiBenedetto, A. Friedman, Hölder estimates for non linear degenerate parabolic systems, Jour. für die Reine und Angewandte Math. 357 (1985), pp. 1–22.

[16] E. DiBenedetto, Chen Ya-zhe, On the local behaviour of solutions of singular parabolic equations, Archive for Rat. Mech. Anal. (to appear).

[17] E. DiBenedetto, Chen Ya-zhe, Boundary estimates for solutions of non linear degenerate parabolic systems, Jour. fúr die Reine und Angewandte Math. (to appear).

[18] L.C. Evans, Application of non linear semigroup theory to certain partial differential equations, in *Non Linear Evolution Equations*, (M.G. Crandall, Editor) (1979).

[19] M.A. Herrero, J.L. Vazquez, Asymptotic behaviour of the solutions of a strongly non linear parabolic problem, Ann. Facultè des Sciences Toulouse, Vol. III (1981), pp. 113–127.

[20] M.A. Herrero, M. Pierre, The Cauchy problem for $u_t = \Delta(u^m)$ when $0 < m < 1$, Trans. AMS Vol. 291, #1 (1985), pp. 145–158.

[21] A. S. Kalashnikov, Cauchy's problem in classes of increasing functions for certain quasi-linear degenerate parabolic equations of the second order, Diff. Uravneniya, Vol. 9 #4 (1973), 682–691.

[22] A. S. Kalashnikov, On uniqueness conditions for the generalized solutions of the Cauchy problem for a class of quasi-linear degenerate parabolic equations, Diff. Uravneniya, Vol. 9 #12 (1973), 2207–2212.

[23] L.I. Kamynin, The existence of solutions of Cauchy problems and boundary-value problems for a second order parabolic equation in unbounded domains:I. Differential Equations Vol. 23 #11 (1987), pp. 1315–1323.

[24] O.A. Ladyzhenskaya, N.A. Solonnikov, N.N. Ural'tzeva, Linear and and quasilinear equations of parabolic type, Trans. Math. AMS, Providence, RI (1968).

[25] O.A. Ladyzenskajia, New equations for the description of motion of viscous incompressible fluids and solvability in the large of boundary value problems for them, Proc. Steklov Inst. Math. #102 (1967), pp. 95–118 (transl. Trudy Math. Inst. Steklov #102 (1967), pp.85–104).

[26] J.L. Lions, Quelques methodes de resolution des problemes a aux limites nonlineaires, Dunod Paris (1969).

[27] L.K. Martinson, K.B. Paplov, Unsteady shear flows of a conducting fluid with a rheological power law, Magnit. Gidrodinamika #2 (1970), pp. 50–58.

[28] L.K. Martinson, K.B. Paplov, The effect of magnetic plasticity in non-Newtonian fluids, Magnit. Gidrodinamika #3 (1969), pp. 69–75.

[29] G. Minty, Monotone (non linear) operators in Hilbert spaces Duke Math. J. #29 (1967), pp. 341–346.

[30] L.E. Payne, G.A. Philippin, Some applications of the maximum principle in the problem of torsional creep, SIAM Jour. Math. Anal. #33 (1977), pp. 446–455.

[31] G.A. Philippin, A minimum principle for the problem of torsional creep, J. Math. Anal. Appl. #68 (1979), pp. 526–535.

[32] M. Pierre, Non linear fast diffusion with measures as data, Proceedings of Non linear parabolic equations: Qualitative properties of solutions, Tesei and Boccardo Eds. Pitman #149 (1985).

[33] M. Pierre, Uniqueness of the solutions of $u_t - \Delta(u)^m = 0$ with initial datum a measure, Non Lin. Anal. TMA, vol. 6 #2 (1982), pp. 175–187.

[34] M. Porzio and P. Vincenzotti, A priori bounds for weak solutions of certain degenerate parabolic equations, (to appear).

[35] E.S. Sabinina, A class of non linear degenerate parabolic equations, Sov. Math. Doklady #143 (1962), pp. 495–498.

[36] S. Tacklind, Nord Acta Regial Soc. Sc. Uppsaliensis Ser.4 10 #3, pp. 3–55 (1936).

[37] A.N. Tychonov, Thèoremes d'unicité pour l'equation de la chaleur Math. Sbornik #42 (1935), pp. 199–216.

[38] L. Veron, Effets régularisants de semigroupes non lineaires dans les éspaces de Banach, Ann. Fac. Sc. Toulouse #1 (1979), pp. 95–103.

[39] D.V. Widder, Positive temperatures in an infinite rod, Trans. AMS #55 (1944), pp. 85–95.

[40] E.DiBenedetto, M.A.Herrero The parabolic p-Laplacian equation. Cauchy problem and initial traces when $1 < p < 2$. (to appear).

For additional references see the Appendix

A NONLINEAR EIGENVALUE PROBLEM INVOLVING FREE BOUNDARIES *

L.A.Peletier

Mathematical Institute, Leiden University
PB 9512 2300 RA Leiden, The Netherlands

In this lecture we consider the eigenvalue problem

$$(I)\begin{cases}(|u|^{m-1}u)'' - \{\lambda - E(x)\}u = 0 \quad -\infty < x < \infty & (1)\\ u \to 0 \quad \text{as} \quad |x| \to \infty, & (2)\end{cases}$$

where $m \geq 1$ and E satisfies the hypotheses

H1. $E \in C(\mathbf{R}) \cap C^1(\mathbf{R} \setminus \{0\})$;
H2. $E(x)$ is an even function;
H3. $E(0) = 1$, $E' < 0$ on $(0, \infty)$ and $\lim_{x\to\infty} E(x) = 0$.

Problem (I) arises in the context of nonlinear diffusion equations of the form

$$u_t = \Delta(u^m) + a(x)u,$$

such as occur for instance in population dynamics [A,GM,N].

For $m = 1$, Problem (I) is well documented and it is known that
- if $\lambda \notin (0,1)$, then there exists no solution of (I);
- if $\lambda \in (0,1)$, then there may or may not exist solutions depending on the function E;
- there exists at most a finite number of solutions;
- the support of any solution equals the entire real line and is thus unbounded.

For $m > 1$ the situation turns out to be quite different. We still have that if $\lambda \notin (0,1)$, then there cannot exist a solution. However, we find that
- if $\lambda \in (0,1)$ then there *always* exists a solution of (I).

In addition we now have
- the support of any solution is a bounded interval.

Thus, let $u(x, \lambda)$ be a solution of (I). Then

$$\xi^*(\lambda) = \sup\{x \in \mathbf{R} : x \in \text{supp } u(\cdot, \lambda)\} < \infty$$

for every $\lambda \in (0,1)$, and similarly for the infimum.

In this case we say that u is a solution of (I) if $(|u|^{m-1}u)'$ exists and is absolutely continuous on $\mathbf{R}$, and u satisfies (1) and (2).

Thanks to the symmetry of E, the eigenfunctions will be either even or odd. For a given $\lambda \in (0,1)$ the principal eigenfunction, which is even and does not change sign on $(-\xi^*, \xi^*)$ was studied in [PT1,2], where its existence, uniqueness and qualitative

---

* This lecture is a report of joint work with with P. de Mottoni (Roma II) and A. Tesei (Roma I)

properties were established. Similar results about the second eigenfunction, which is odd and only changes sign at $x = 0$ were obtained in [MPT].

In this lecture we present some of the ideas underlying the proofs of these results, discuss the location of the free boundaries, their dependence on λ and make some observations about higher eigenfunctions.

We shall mainly look at odd eigenfunctions and thus it will suffice to consider (1) on the half line $(0, \infty)$ and impose a zero Dirichlet boundary condition at $x = 0$:

$$u(0) = 0. \tag{3}$$

The analysis of the even eigenfunctions is entirely analogous.

We fix $\lambda \in \mathbf{R}$ and set

$$|u|^{m-1}u = y, \quad m^{-1} = \beta, \quad a(x) = \lambda - E(x).$$

The analysis of Problem (1-3) is based on a careful study of the initial value problen

$$\begin{cases} y'' - a(x)y^{\beta} = 0, & y > 0 \quad \text{for} \quad x > 0 \\ y(0) = 0, & y'(0) = \omega. \end{cases}$$

For every $\omega > 0$, Problem (II) has a unique solution $y(x, \omega)$ on $[0, \sigma]$, where

$$\sigma = \sup\{x > 0 : y(\cdot, \omega) > 0\}.$$

We shall say that $\omega \in \Sigma$ if either $\sigma < \infty$ or $y(x, \omega) \to 0$ as $x \to \infty$, and that $\omega \in \Sigma_0$ if $\omega \in \Sigma$ and in addition $y(\cdot, \omega) \in C^1([0, \infty))$. Plainly, $\omega \in \Sigma_0$ if anf only if the function

$$u(x) = \{y(x, \omega)\}^{\beta}$$

is a solution of (1-3).

As a first observation we note that

$$\lambda \geq 1 \Rightarrow a(x) > 0 \Rightarrow y'' > 0,$$

which means that $y(x, \omega) \geq \omega x$ and hence the set Σ is empty. On the other hand,

$$\lambda \leq 0 \Rightarrow a(x) < 0 \Rightarrow y'' < 0,$$

and so either $\sigma < \infty$ or $\sigma = \infty$ and $y'(x, \omega) > 0$ for all $x > 0$. Thus, in this case the set Σ_0 is empty.

If $\lambda \in (0, 1)$ there exists a unique $\alpha \in (0, \infty)$ such that

$$a(\alpha) = \lambda - E(\alpha) = 0$$

and

$$a(x) \begin{cases} < 0 & \text{on} \quad [0, \alpha) \\ > 0 & \text{on} \quad (\alpha, \infty) \end{cases}$$

so that

$$y''(x)\begin{cases} <0 & \text{on} \quad (0,\alpha) \\ >0 & \text{on} \quad (\alpha,\infty). \end{cases}$$

Thus, if $\omega \in \Sigma$ and $\sigma > \alpha$, then $y'(\bar{\alpha},\omega) < 0$ for $\alpha \le \bar{\alpha} < \sigma$. Writing

$$\omega^- = \inf\{\omega > 0 : \sigma > \alpha)$$

and

$$\omega^+ = \sup\{\omega > \omega^- : y'(\alpha,\omega) < 0\},$$

it can be shown that $y'(\alpha,\omega) < 0$ for all $\omega \in [\omega^-,\omega^+)$.

The main result for Problem (II) is formulated in terms of the function $\xi : (0,\omega^+) \to \mathbf{R}^+$ defined by

$$\xi(\omega) = \begin{cases} \sigma & \text{if} \quad 0 < \omega \le \omega^- \\ \sup\{x \in (\alpha,\sigma) : y' < 0 \text{ on } (\alpha,x)\} & \text{if} \quad \omega^- < \omega < \omega^+. \end{cases}$$

Theorem 1. *There exists a unique initial slope* $\omega^* \in (\omega^-,\omega^+)$ *such that*
(a) $\xi \in C([0,\omega^+]) \cap C^1((0,\omega^*) \cup (\omega^*,\omega^+))$;
(b) ξ *is strictly increasing on* $(0,\omega^*)$ *and*

$$\xi(\omega) = O(\omega^{(1-\beta)/(1+\beta)}) \quad \text{as} \quad \omega \downarrow 0;$$

(c) ξ *is strictly decreasing on* (ω^*,ω^+) *and*

$$\xi(\omega) - \alpha = O((\omega^+ - \omega)^{1/2}) \quad \text{as} \quad \omega \uparrow \omega^+.$$

The graph of $\xi(\omega)$ can be thought of as a bifurcation diagram for solutions of Problem (II) (See Fig. 1). The cusp–like behaviour of $\xi(\omega)$ near ω^* was analysed in [PT2].

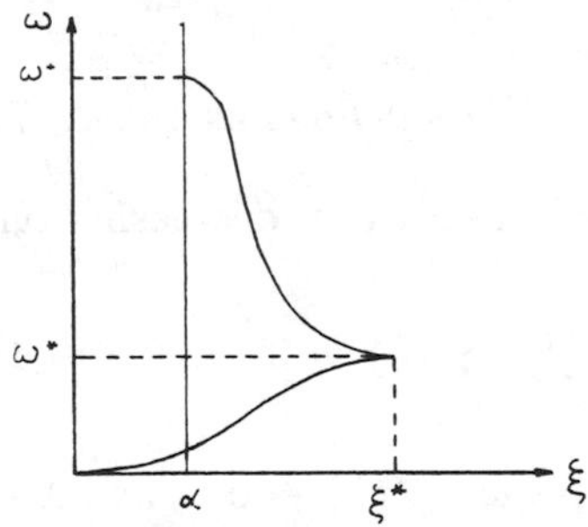

Fig. 1. Bifurcation diagram for Problem (II).

Because when $\omega \in (0,\omega^*)$, y vanishes at $x = \xi$ and when $\omega \in (\omega^*,\omega^+)$, y' vanishes at $x = \xi$, we refer to the lower part of the graph as the *Dirichlet branch* and the higher part as the *Neumann branch*. The function $y(\alpha,\omega^*)$ is called a *free boundary solution* with *free boundary* ξ^*.

We now return to the eigenvalue Problem (1-3) which we write as

$$\text{(III)}\begin{cases} y'' - \{\lambda - E(\alpha)\}y^\beta = 0, & y \geq 0 \\ y(0) = 0,\ y(x) \to 0 \quad \text{as} & x \to \infty. \end{cases}$$

In the previous analysis we found that if $\lambda \notin (0,1)$ then (III) has no solutions, so we take $\lambda \in (0,1)$. By Theorem 1, there exists for each $\lambda \in (0,1)$ a uniform slope $\omega^*(\lambda)$ such that

$$y(\alpha,\omega^*) > 0 \quad \text{on} \quad (0,\xi^*)$$
$$y(\xi^*,\omega^*) = y'(\xi^*,\omega^*) = 0,$$

where $\xi^*(\lambda) = \xi(\omega^*(\lambda))$.

To analyse the functions of ω^* and ξ^*, we return to Problem (II), but now we vary λ rather than ω.

Lemma 2. [MPT] *Let $y(\cdot,\omega,\lambda)$ be the solution of Problem (II). Then*

$$\lambda_1 < \lambda_2 \Rightarrow y(\alpha,\omega,\lambda_1) < y(\alpha,\omega,\lambda_2)$$

whenever both solutions are defined and nonnegative.

Fix $\lambda = \lambda_1$ and choose $\omega = \omega^*(\lambda_1)$. Then, by Lemma 2 raising λ to $\lambda_2 > \lambda_1$, we move to the Neumann branch of the bifurcation diagram (See Fig.1) and we have to lower ω to return to a solution with a free boundary, i.e.

$$\lambda_1 < \lambda_2 \Rightarrow \omega^*(\lambda_1) > \omega^*(\lambda_2).$$

By a more delicate analysis one can also show that

$$\lambda_1 < \lambda_2 \Rightarrow \xi^*(\lambda_1) > \xi^*(\lambda_2).$$

Summarizing we have

Theorem 3. *Let $\omega^*(\lambda)$ and $\xi^*(\lambda)$ be defined as above. Then*

(a) ω^* *and* ξ^* *are strictly decreasing on* $(0,1)$;

(b) $$\omega^*(\lambda),\ \xi^*(\lambda) \to 0 \quad \text{as} \quad \lambda \to 1;$$

(c) $$\omega^*(\lambda) \to \hat{\omega},\ \xi^*(\lambda) \to \infty \quad \text{as} \quad \lambda \to 0,$$

where

$$\hat{\omega} < \infty \quad \textit{if} \quad \int_0^\infty x^\beta E(x)\,dx < \infty.$$

We conclude with a few observations about higher eigenfunctions of Problem (1-3). For any $\lambda \in (0,1)$ and every $k \in \mathbf{N}$ we can seek an eigenfunction with precisely k zeros and compact support $[-\xi_k^*, \xi_k^*]$. For $k = 0$ the existence and uniqueness of such an eigenfunction was proved in [PT2] and for $k = 1$ the same was proved in [MPT]. By considering the auxiliary problem

$$\begin{cases} y'' - a(\alpha)y^\beta = 0, y > 0 & \text{for} \quad x > \theta \\ y(\theta) = 0, \; y'(\theta) = \omega, \end{cases}$$

where $\theta \in (0, \alpha)$ and $\omega > 0$ we can define functions $\omega^*(\lambda, \theta)$ and $\xi^*(\lambda, \theta)$. It is clear that

$$\begin{cases} \omega^*(\lambda,\theta) \to \omega^*(\lambda), & \xi^*(\lambda,\theta) \to \xi^*(\lambda) \quad \text{as} \quad \theta \to 0 \\ \omega^*(\lambda,\theta) \to 0, & \xi^*(\lambda,\theta) \to \alpha \quad \text{as} \quad \theta \to \alpha. \end{cases}$$

On the other hand, as ω increases from 0 to ω^-, $\xi(\omega)$ moves from 0 to α and $y'(\xi(\omega), \omega)$ from 0 to $y'(\alpha, \omega)$. Hence, since ξ is strictly increasing on $(0, \omega^-)$ by Theorem 1, we can define the function

$$\phi(\theta) = -y'(\theta, \xi^{-1}(\theta)).$$

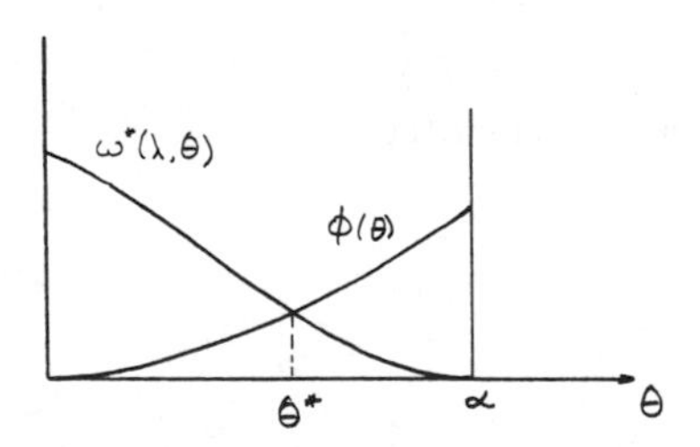

Fig. 2. The functions $\omega^*(\lambda, \theta)$ and $\phi(\theta)$.

Thus by the continuity of $\omega^*(\lambda, \cdot)$ and ϕ there exists a point $\theta^* \in (0, \alpha)$ such that

$$\omega^*(\lambda, \theta^*) = \phi(\theta^*),$$

ensuring the existence of the solution u_3. The existence of u_2 can similarly be proved and it is conjectured that in fact for any $k \in \mathbf{N}$ it is possible to find an eigenfunction u_k.

By a comparison argument it is possible to show that

$$\xi_0^* > \xi_1^* > \xi_2^* > \xi_3^*$$

and it is conjectured that the ξ_k^* form a monotone sequence.

References

[A] Aronson, D.G., *Bifurcation phenomena associated with non–linear diffusion mechanisms*, in "Partial Differential Equations and Dynamical Systems" (Ed. W.E. Fitzgibbon, III), Pitman, New York, 1984.

[GM] Gurtin, M.E., & R.C. MacCamy, *On the diffusion of biological populations.* Math. Biosci. **33** (1977), 35–49.

[MPT] Mottoni, P. de, L.A. Peletier & A. Tesei, *Free boundary eigenfunctions for a nonlinear degenerate eigenvalue problem.* J. reine angew. Math. **394** (1989), 31–58.

[N] Namba, T. *Density–dependent dispersal and spatial distribution of a population.* J. Theor. Biol. **86** (1980), 351–363.

[PT1] Peletier, L.A. & A. Tesei, *Diffusion in homogeneous media: localisation and positivity.* Ann. Mat. Pura Appl.**141** (1985), 307–330.

[PT2] Peletier, L.A. & A. Tesei, *Global bifurcation of stationary solutions of degenerate diffusion equations.* Adv. Appl. Math. **7** (1986), 435–454.

For additional references see the Appendix

TWO NONLINEAR DIFFUSION EQUATIONS WITH FINITE SPEED OF PROPAGATION

JUAN LUIS VAZQUEZ

Dpto. de Matemáticas
Univ. Autónoma de Madrid
28049 Madrid, Spain.

For J. Hale on his 60th birthday

1. In this lecture I will consider two models of nonlinear diffusion equation which have received a great deal of attention from mathematicians in recent times, namely the *porous medium equation*

$$u_t = \Delta(u^m), \quad m > 1, \quad x \in \mathbf{R}^N, t > 0 \tag{1}$$

and the *p-Laplacian equation*

$$u_t = \operatorname{div}(|\nabla u|^{p-2} \nabla u), \quad p > 2\,. \tag{2}$$

In presenting an account of progress done in the last years I will put some emphasis on the property of propagation with finite speed in which I have been particularly interested, and I will also stress the deep analogy existing in these respects between two equations which are in principle very different.

Both (1) and (2) are examples of degenerate parabolic equation. Thus, if we view equation (1) in terms of Fickian Diffusion, we discover that the diffusivity $D(u) = mu^{m-1}$ vanishes at the level $u = 0$. This has a series of consequences which strongly depart from what is typically expected from a parabolic equation. Most prominent among these consequences is the Finite Propagation Property, first established rigorously by Oleinik and her collaborators around 1958, cf. [OKC], which states that disturbances from the level $u = 0$ propagate with finite speed and which is technically phrased as follows: if $u(x,0)$ has compact support, so does $u(x,t)$ for every $t > 0$ (however, the support grows eventually to cover the whole spatial domain). Properties of equation (1) related to finite propagation are as a rule more easily described in terms of the variable

$$v = \tfrac{m}{m-1} u^{m-1}, \tag{3}$$

which represents (up to a constant) the *pressure* of the fluid in the description of the flow of gas through a porous medium, u standing for the density and $w = -v_x$ for the velocity ($v_x = \nabla v$ denotes the spatial gradient of v). In terms of v equation (1) becomes

$$v_t = (m-1)\, v\, \Delta v + |v_x|^2. \tag{4}$$

We notice at once that for $v \approx 0$ a good approximation to (4) should be the equation

$$v_t = |v_x|^2 \tag{5}$$

which is a *Hamilton-Jacobi* equation.The equation satisfied by w is (in one space dimension to make things simpler)

$$w_t + (w^2)_x = (m-1)(v\, w_x)_x\,, \tag{6}$$

to be approximated by the *conservation law*

$$w_t + (w^2)_x = 0. \tag{7}$$

It is well known that equations like (7) develop *shocks* : along certain lines $x = s(t)$ the solution w exhibits jump discontinuities governed by the so called *Rankine-Hugoniot* conditions, which take in this case the form $[w]\, s'(t) = [w^2]$, where [.] denotes jump, see [L] or [S]. We will find similar things to happen for the solutions of (1). An important particular type of solution of (1) on which this is easily visualized is the so called *linear pressure wave* given by

$$v_c(x,t) = c(ct - x_1)_+, \quad x = (x_1, \cdots, x_N) \tag{8}$$

Where $(.)_+$ means max(.,0). Since $\Delta v_c = 0$, (8) is at the same time a solution of (4) for any m and a solution of (5). Let now $N = 1$. Then $w = -v_x$ is not only a solution of (6) but also a solution of (7). In fact, it is a viscosity solution satisfying the correct entropy conditions.We also observe that both v_x and v_t are discontinuous over the line $x = ct$ which separates the regions $v_c > 0$ and $v_c = 0$. Such a line is called an *interface* or *free boundary* of the solution. For $N > 1$ we have to replace the word line by hyperplane in $\mathbf{R}^N \times \mathbf{R}$.

Another interesting set of explicit solutions of (1) is given by the formula ([B],[ZK])

$$U(x,t) = t^{-k}\left(C - \frac{k(m-1)}{2mN}\cdot\frac{|x|^2}{t^{2k/N}}\right)_+^{1/(m-1)}, \tag{9}$$

where $k = (m - 1 + 2/N)^{-1}$ and $C > 0$ is an arbitrary constant which can be uniquely determined by fixing the *total mass* $M = \int U(x,t)\, dx$. This solution takes on initial data

$$U(x,0) = M\delta(x), \tag{10}$$

(δ is Dirac's delta function) hence it can be called a *fundamental solution* of (1) by analogy with the terminology used for linear equations; in fact, the solution is usually known as *source-type* solution or Barenblatt's solution. Its positivity set $\mathcal{P} = \{(x,t) \in \mathbf{R}^N \times (0,\infty) : U(x,t) > 0\}$ is the region of (x,t)-plane contained inside the hypersurface

$$|x| = c(m,N)(M^{m-1}t)^{k/N} \equiv r_M(t). \tag{11}$$

In $\mathcal{P}$ our solution U is C^∞ smooth and satisfies (1) in a classical sense. But on the interface $|x| = r(t)$ the derivatives of the pressure V, i.e. V_x and V_t, exhibit jump discontinuities and (their lateral limits) satisfy the equation $V_t = |V_x|^2$. Moreover, the velocity of the interface is related to the jumps, namely $r'(t) = [V_x] \equiv V_{x,out} - V_{x,in}$, where $V_{x,out} = 0$ and $V_{x,in}$ is the limit of V_x as (x,t) approaches the interface with $(x,t) \in \mathcal{P}$. This is the Rankine-Hugoniot condition. Finally we remark that $r(t)$ is C^∞ for $t > 0$.

These considerations motivate the following program for the investigation of equation (1) :

(I) Solutions need not be classical, hence we have to consider a suitable class of weak solutions. Prove existence, uniqueness and partial regularity for these weak solutions. The restiction to nonnegative solutions is acceptable since it turns out in most of the applications.

(II) There is finite speed of propagation, hence interfaces appear whenever the initial data vanish in some region; the study of interfaces is of primary importance.

(III) The behaviour of the solution near the interface is governed by the linear pressure wave (8) and the approximations (5), (7) in a small scale.

(IV) On the contrary, near a point (x,t) where $u(x,t) > 0$ (1) will behave like a parabolic equation, producing for instance C^∞ smoothness.

(V) The asymptotic behaviour is governed by the fundamental solutions (9).

A theory should give a precise description of these behaviours.

Much of this program has been carried out in recent years, specially for the class of nonnegative solutions : for a general survey of results up to 1986 the reader can consult [A]. A detailed discusion of the relationship between (1) and (5) and (7) can be found in [V2], see also [V1]. Recent results on interface behaviour are contained in [CVW] for $N > 1$, and [V3] for $N = 1$. In these papers further references are given.

Let us briefly discuss one aspect of the theory. It is a standard practice in P.D.E. to base a theory on a number of a priori estimates combined with several other tools. In the case of equation (1) and for solutions $u \geq 0$ the crucial estimate seems to be the following lower bound

$$\Delta v \geq -\frac{k}{t}, \quad k = (m - 1 + \tfrac{2}{N})^{-1} \tag{12}$$

due to Aronson and Bénilan [AB]. In terms of w (12) reads $\operatorname{div}(w) \leq k/t$. For $N = 1$ this coincides with Oleinik's entropy criterion [O] to select good solutions of conservation laws like (7), a connection that is at least remarkable. For a detailed discussion of this aspect see [AV] and also [LSV]. Note that (12) is an absolute bound in the sense that it does not involve any particular information about the solution for which it holds.

A typical result where (12) is essentially used is the following description of an interface in $N = 1$, say the right-hand one given by

$$s(t) = \sup\{x : u(x,t) > 0\}, \tag{13}$$

when the initial data $u(x,0) \geq 0$ have compact support. There is a time $t^* \geq 0$ such that $s(t) = s(0)$ for $0 \leq t \leq t^*$ and s is strictly increasing for $t > t^*$. Moreover, $s \in C^\infty(t^*, \infty)$,

$$s'(t) = -v_x(s(t)-, t) > 0 \tag{14}$$

and

$$s'' + \frac{m}{(m+1)t} s'(t) \geq 0. \tag{15}$$

Finally $s(t) = r_M(t) + x_0 + o(1)$ and $s'(t) = r'_M(t) + o(1/t)$ as $t \to \infty$. $r_M(t)$ is the expression for the radius of the support of the fundamental solution (9) with the same mass M as u and x_0 is the center of mass $M^{-1}\int x u_0(x)dx$, an invariant of the evolution. Cf. the reference [A] for details.

Corresponding results for $N > 1$ are much more difficult to prove and only a partial picture exists. Thus, it is known that solutions and interfaces are Hölder continuous [CF], even Lipschitz continuous for large times [CVW], but the expected C^∞ result has not been obtained to our knowledge.

2. In principle the p-Laplacian equation offers more difficulties, since it replaces the linear Δ operator of equation (1) by an essentially nonlinear differential operator. This fact without doubt causes some different properties, along with extra technical difficulties that will be appreciated in E. di Benedetto's contribution and appear of course when we try to carry out a program as that outlined above for (1). It is however remarkable (and surprising for many a researcher) that such a program can be developped along the same lines, giving very similar results. Paramount to this approach is the identification of the correct variables. Thus the "pressure" is now defined as

$$v = \tfrac{p-1}{p-2}\, u^{\frac{p-2}{p-1}}\,. \tag{16}$$

We then obtain from (2) the equation for v

$$v_t = \tfrac{p-2}{p-1}\, v \operatorname{div}(|\nabla v|^{p-2}\nabla v) + |\nabla v|^p \tag{17}$$

which is to be compared near $v = 0$ with

$$v_t = |\nabla v|^p\,. \tag{18}$$

The corresponding velocity w is defined as

$$w = -|\nabla v|^{p-2}\nabla v\,. \tag{19}$$

With these definitions both (1) and (2) can be written in the form of mass balance

$$u_t + \operatorname{div}(uw) = 0, \tag{20}$$

which identifies w as a point velocity. Differentiation of (17) gives for $N = 1$ the equation for $z = -v_x$

$$z_t + (|z|^p)_x = (p-2)(v|z|^{p-2}z_x)_x\,, \tag{21}$$

which is to be approximated by the conservation law

$$z_t + (|z|^p)_x = 0\,. \tag{22}$$

The equation for the velocity is

$$w_t + \tfrac{p}{2}(w^2)_x = (p-2)|w|^{\frac{p-2}{p-1}}(v\, w_x)_x\,.$$

Again we obtain a model for interface behaviour in the form of a linear pressure wave

$$v_c(x,t) = (|c|^p t - cx)_+, \quad c \in \mathbf{R}, \tag{23}$$

which solves at the same time (17) for all p and (18). Here the RH condition on the interface $|x| = |c|^{p-2}ct$ is

$$x'(t) = -|v_{x,in}|^{p-2} v_{x,in} \tag{24}$$

which agrees with (22) ($v_{x,in}$ denotes the lateral value of v_x at the interface from the side of the support).

Let us now turn to the fundamental solutions in dimension $N \geq 1$. Analogously to (9) we can easily check that the following functions are weak solutions of (2) :

$$U(x,t) = t^{-k}\left(C - q|\xi|^{\frac{p}{p-1}}\right)_+^{\frac{p-1}{p-2}} \tag{25}$$

where

$$\xi = xt^{-k/N}, \; q = \tfrac{p-2}{p}(\tfrac{k}{N})^{\frac{1}{p-1}},$$

$C > 0$ is related to the total mass M and k is now defined to be $(p-2+p/N)^{-1}$. Again $U(x,t)$ takes on as initial data $M\delta(x)$, hence it is what we call a fundamental solution. The discussion of the solutions (9) applies here mutatis mutandis; thus, we find that the equation $v_t = |v_x|^p$ holds at the interface $|x| = r(t) \equiv c(p,N)(M^{p-2}t)^{k/N}$, that v_x and v_t jump at the interface and are continuous in the positivity set $\mathcal{P}$. The solution is not C^∞ smooth in $\mathcal{P}$ however, since the equation degenerates at all points where $u_x = 0$ (recall that the diffusivity is given by $D(u) = |u_x|^{p-2}$). This applies to U at $x = 0$.

An attempt to develop a theory for the Cauchy problem for the p-Laplacian equation in parallel to the corresponding theory for the porous medium equation has been done in one space dimension in [EV1]. As a matter of fact, this paper considers the equation

$$u_t = (u^r|u_x|^{p-2}u_x)_x \tag{26}$$

for all $r > 0, p > 1$, which includes both equations (1) and (2) as particular cases. The solutions are as usual nonnegative and it is also assumed that they have finite mass: $\int u(x,t)\,dx < \infty$. Finite propagation occurs for $r + p > 2$. We define the pressure as

$$v = \tfrac{p-1}{r+p-2}\, u^{\frac{p+r-2}{p-1}} \tag{27}$$

and find a *crucial* estimate of the form

$$(|v_x|^{p-2}v_x)_x \geq -\frac{1}{(2p+r-2)t}, \tag{28}$$

which generalizes (12). This is used in [EV1] to show the existence of strong solutions for the Cauchy problem, and to describe the interface behaviour and the large time behaviour for solutions with compact support (in the x variable). Thus, the right-hand interface $s(t)$, defined as in (13) has a waiting time $t^* \geq 0$ and is strictly increasing and smooth afterwards, with an equation

$$s'(t) = -(|v_x|^{p-2}v_x)(s(t)-,t) \tag{29}$$

and a semiconvexity inequality

$$s''(t) + \left(\tfrac{2p+r-3}{2p+r-2}\right)\frac{s'(t)}{t} \geq 0. \tag{30}$$

Besides, for $t > t^*$ v_x and v_t are continuous on each side of the interface and up to it, with a jump discontinuity precisely across the interface, on which the equation $v_t = |v_x|^p$ is satisfied.

Regarding the large-time behaviour, as in the case of equation (1) $s(t)$ converges to $r_M(t) + x_0$ with error $o(1)$, the difference being that now the center of mass is not necessarily invariant in time and x_0 denotes rather the asymptotic center of mass.

Corresponding results in several space dimensions are as usual more difficult to prove. We have recently obtained [EV2] an estimate

$$\operatorname{div}(|\nabla v|^{p-2}\nabla v) \geq -\frac{C}{t} \tag{31}$$

with $C = C(p, r, N) > 0$, which holds in the parameter range $r + p - 2 + p/N > 0$ (do not complain, please, about another seemingly technical and awful condition; it is in fact necessary). Estimate (31) should be a basic step in our approach to studying (2) for $N > 1$.

3. The third part of this lecture will be devoted to comment on the description of the asymptotic behaviour of solutions of the Cauchy problem in terms of the above described fundamental solutions, a kind of result where the similarity between (1) and (2) is very apparent.

Let us begin with the porous medium equation. The following three results represent basic information about the large-time behaviour and also relate this behaviour to the classification of solutions with a singularity at $(0,0)$.

THEOREM 1. *A nonnegative solution of equation* (1) *in* $Q = \mathbf{R}^N \times (0,\infty)$ *which takes on initial data*

$$u(x,0) = 0 \quad for \quad x \neq 0 \tag{32}$$

is necessarily one of the solutions $U(x,t;M)$ *given by* (9) *unless* $u \equiv 0$.

THEOREM 2. *Let* u *be a nonnegative solution of* (1) *in* Q *such that* $\int u(x,t)\,dx = M < \infty$ *(the mass is an invariant). Then*

$$\lim_{t\to\infty} t^k |u(x,t) - U(x,t;M)| = 0 \tag{33}$$

uniformly in $x \in \mathbf{R}^N$. *Here* $k = (m - 1 + \frac{2}{N})^{-1}$.

The case where u has changing sign has been less studied. One replaces equation (1) by its natural generalization

(1')
$$u_t = \Delta(|u|^{m-1}u).$$

The following result shows the eventual behaviour of solutions to (1').

THEOREM 3. *Let u be any solution of* (1) *in Q such that $\int u(x,t)\,dx = M > 0$. Then there exists a time $T < \infty$ such that $u(x,t) \geq 0$ for every $t \geq T$ and $x \in \mathbf{R}^N$.*

Kamin [K] established Theorem 2 in 1973 for $N = 1$ and data with compact support. A proof valid for $N > 1$ and $u_0 \in L^1(\mathbf{R}^N)$ is given in [FK], 1980. As for Theorem 1 it is known that every solution of (1) has an initial trace which is a measure [AC], 1983, and nonegative solutions with such initial data are unique [DK], 1984. Theorem 3 comes just out of the oven, [KV2].

The three theorems are true verbatim for the p-Laplacian equation if only one interprets $U(x,t)$ as the function given in (27) and takes $k = (p - 2 + p/N)^{-1}$. Theorems 1 and 2 are contained in a paper to appear with Kamin [KV1] and Theorem 3 in [KV2]. As shown in the papers the same line of proof works for both equations (subject of course to technical differences).

4. I will end this exposé with a glimpse on what happens in the range of exponents where finite propagation does not hold, i.e. $m \leq 1$ for the porous medium equation and $p \leq 2$ for the p-Laplacian equation, an area in which progress is very recent.

The limit cases, $m = 1$ for (1) and $p = 2$ for (2), are just the classical heat equation, whose properties are well known.

Let us take then equation (1) for $m < 1$. In the range $0 < m < 1$ for every $u_0 \in L^1_{loc}(\mathbf{R}^N), u_0 \geq 0$, there exists a solution $u \in C^\infty(Q), u > 0$ everywhere, cf. [AB],[HP]. Hence no interfaces appear, we have *fast diffusion*. For $m > 1 - \frac{2}{N}$ and solutions with finite mass the fundamental solutions defined as in (9) (but now positive everywhere) still give the asymptotic behaviour and Theorem 2 above holds, [KF], [KV2]. However Theorem 3 does not hold, [KV2]. Fundamental solutions do not exist for $m < 1 - \frac{2}{N}$, [BF].

Cases with $m \leq 0$ have been studied in one space dimension, the equation being now written in the form

(1")
$$u_t = (u^{m-1}u_x)_x$$

to preserve its parabolicity. In all cases $m < 1$ the level $u = 0$ is singular, since $D(u) = u^{m-1} \to \infty$ as $u \to 0$. This singularity forces us to work with nonnegative solutions. For $0 \geq m > -1$ a theory of positive , C^∞ solutions with $L^1_{loc}(\mathbf{R})$ initial data can be developped much as for $m > 0$ [ERV] but *uniqueness fails.* We can even prescribe arbitrary "Neumann data at infinity"

(34)
$$\lim_{x\to\infty} u^{m-1}u_x = -f(t)$$

(35)
$$\lim_{x\to-\infty} u^{m-1}u_x = g(t)$$

with $f, g \in L^\infty_{loc}(\mathbf{R})$ (for instance) and $f, g \geq 0$ (this is essential) and obtain a unique solution in some interval $0 < t < T$ (possibly $T < \infty$ and we have then extinction in finite time). The complete development of this surprisingly well-posed problem can be found in [RV].

Finally, things are again different for $m \leq -1$. If we consider equation (1") with integrable initial data, then no solution exists, though solutions exist for nonintegrable data, [H1], [H2]. The nonexistence result is explained in the limit case $m = -1$ in [ERV] as a boundary layer phenomenon: we consider the maximal solution u_m of (1) for $m > -1$ with $u_m(x, 0) = u_0(x) \in L^1(\mathbf{R})$, $u_0 \geq 0$ and let $m \to -1$. Then for every $\epsilon > 0$, $u_m(x,t) < \epsilon$ for $t \geq t_{\epsilon,m}$ and $t_{\epsilon,m} = O(m+1)$ for fixed ϵ. Thus, as $m \to -1$ diffusion becomes so fast that the solution just disappears in becoming homogeneously distributed over the whole space.

Corresponding results for the p-Laplacian equation should be similar broadly speaking, but complete details lack.

5. This very brief presentation omits many interesting developments, for which I can only offer the excuse of lack of space in a lecture format. Only to mention a couple of examples, nothing is said about mixed initial and boundary-value problems, for which much is known, see for instance [AP], or about the approach to $m < 1$ through a pressure equation similar to (4) as done in [BPU].

REFERENCES

[A] D.G.ARONSON, *The Porous Medium Equation*, in Some Problems in Nonlinear Diffusion (A. Fasano & M.Primicerio eds.), Lecture Notes in Maths. 1224, Springer, New York, 1986.

[AB] D.G.ARONSON & P.BENILAN, *Régularité des solutions de l'équation des milieux poreux dans* $\mathbf{R}^N$, Comptes Rendus Ac. Sci. Paris **288** (1979), 103-105.

[AC] D.G.ARONSON & L.A.CAFFARELLI, *The initial trace of a solution of the porous medium equation*, Trans. Amer. Math. Soc. **280** (1983), 351-366.

[AP] D.G.ARONSON & L.A.PELETIER, *Large-time behaviour of solutions of the porous media equation in bounded domains*, J. Diff. Equations **39** (1981), 378-412.

[AV] D.G.ARONSON & J.L.VAZQUEZ, *The porous medium equation as a finite-speed approximation to a Hamilton-Jacobi equation*, J. d'Analyse Nonlinéaire (Ann. Inst. H. Poincaré) **4** (1987), 203-230.

[B] G.I.BARENBLATT, *On some unsteady motions of a liquid or a gas in a porous medium*, Prikl. Mat. Mekh. **16** (1952), 67-78 (in Russian).

[BPU] M.BERTSCH, R. dal PASSO & M.UGHI, *Discontinuous viscosity solutions of a degenerate parabolic equation*, to appear.

[BF] H.BREZIS & A.FRIEDMAN, *Nonlinear parabolic equations involving measures as initial data*, J.Math. Pures Appl. **62** (1983), 73-97.

[CVW] L.A.CAFFARELLI, J.L.VAZQUEZ & N.I.WOLANSKI, *Lipschitz continuity of solutions and interfaces to the N-dimensional porous medium equation*, Indiana Univ. Math. J. **36** (1987), 373-401.

[CF] L.A.CAFFARELLI & A.FRIEDMAN, *Regularity of the free boundary of a gas flow in an n-dimensional porous medium*, Indiana Univ. Math. J. **29** (1987), 361-391.

[DK] B.E.J.DAHLBERG & C.E.KENIG, *Nonnegative solutions of the porous medium equation*, Comm. Partial Diff. Eq. **9** (1984), 409-437.

[ERV] J.R.ESTEBAN, A.RODRIGUEZ & J.L.VAZQUEZ, *A nonlinear heat equation with singular diffusivity*, Comm. Partial Diff. Eq. **13** (1988), 985-1039.

[EV1] J.R.ESTEBAN & J.L.VAZQUEZ, *Homogeneous diffusion in* $\mathbf{R}$ *with power-like nonlinear diffusivity*, Archive Rat. Mech. Analysis **103** (1988), 39-80.

[EV2] J.R.ESTEBAN & J.L.VAZQUEZ, work in preparation.

[FK] A.FRIEDMAN & S.KAMIN, *The asymptotic behaviour of a gas in an n-dimensional porous medium*, Trans. Amer. Math. Soc. **262** (1980), 551-563.

[H1] M.A.HERRERO, *A limit case in nonlinear diffusion*, Nonlinear Anal., to appear.

[H2] M.A.HERRERO, *Singular diffusion on the line*, to appear.

[HP] M.A.HERRERO & M.PIERRE, *The Cauchy problem for* $u_t = \Delta(u^m)$ *when* $0 < m < 1$, Trans. Amer. Math. Soc. **291** (1985), 145-158.

[K] S.KAMIN, *The asymptotic behaviour of the solution of the filtration equation*, Israel J. Math. **14** (1973), 76-87.

[KV1] S.KAMIN & J.L.VAZQUEZ, *Fundamental solutions and asymptotic behaviour for the p-Laplacian equation*, Rev. Mat. Iberoamericana, to appear.

[KV2] S.KAMIN & J.L.VAZQUEZ, *Asymptotic behaviour of solutions of the porous medium equation with changing sign*, preprint.

[L] P.LAX, *Hyperbolic systems of conservation laws II*, Comm. Pure Appl. Math **10** (1957), 537-566.

[LSV] P.L.LIONS, P.E.SOUGANIDIS & J.L.VAZQUEZ, *The relation between the porous medium equation and the eikonal equation in several space variables*, Rev. Mat. Iberoamer. **3** (1987), 275-310.

[O] O.A.OLEINIK, *Discontinuous solutions of nonlinear differential equations*, Amer. Math. Soc. Translations **(2) 26** (1983), 95-172.

[OKC] O.K.OLEINIK, A.S.KALASHNIKOV & Y.L.CZHOU, *The Cauchy problem and boundary value problems for equations of the type of nonstationary filtration*, Izv. Akad. Nauk SSSR, Ser. Mat. **22** (1958), 667-704 (in Russian).

[RV] A.RODRIGUEZ & J.L.VAZQUEZ, *A well-posed problem in singular Fickian diffusion*, to appear.

[S] J.SMOLLER, Shock Waves and Reaction-Diffusion Equations, Springer Verlag, New York, 1983.

[V1] J.L.VAZQUEZ, *Behaviour of the velocity of one-dimensional flows in porous media*, Trans. Amer. Math. Soc. **286** (1984), 787-802.

[V2] J.L.VAZQUEZ, *Hyperbolic aspects in the theory of the porous medium equation*, in Metastability and Incompletely Posed Problems (S. Antman et al. eds.), Springer Verlag, New York, 1987, 325-342.

[V3] J.L.VAZQUEZ, *Regularity of solutions and interfaces of the porous medium equation via local estimates*, Proc. Royal Soc. Edinburgh **112A** (1989) (to appear).

[ZK] Y.B.ZEL'DOVICH & A.C.KOMPANYEETS, *On the theory of heat conduction depending on temperature*, Lectures dedicated on the 70th anniversary of A. F. Joffe, Akad. Nauk SSSR, 1950, (in Russian).

Appendix: Additional References

M. Gurtin:

Gurtin, M.E., Multiphase thermomechanics with interfacial structure. 1. Heat conduction and the capillary balance law, Archive for Rational Mechanics and Analysis, **104**, 195–221 (1988).

Angenent, S. and Gurtin, M.E., Multiphase thermomechanics with interfacial structure. 2. Evolution of an isothermal interface, Archive for Rational Mechanics and Analysis, **108**, 323–391 (1998).

Gurtin, M.E., A mechanical theory for crystallization of a rigid solid in a liquid melt: melting–freezing waves, Archive for Rational Mechanics and Analysis. Forthcoming.

Gurtin, M.E., and Podio–Guidugli, P., Hyperbolic theory for the evolution of plane curves. SIAM Journal on Mathematical Analysis. Submitted.

M. Renardy:

Renardy, M., Existence of steady flows of viscoelastic fluids of Jeffreys type with traction boundary conditions, Diff. Integral Eq. **2**, 431–437 (1989).

Renardy, M., Inflow boundary conditions for steady flow of viscoelastic fluids with differential constitutive laws, Rocky Mt. J. Math. **19** (1989).

D. Joseph:

Joseph, D.D., *Fluid Dynamics of Viscoelastic Liquids*, Applied Math. Sciences, Springer–Verlag 1990, to appear.

J.A. Nohel:

Malkus, D.S., Nohel, J.A., and Plohr, B.J., Analysis of new phenomena in shear flow of Newtonian fluids, SIAM Journal of Applied Math. (1988).

Malkus, D.S., Nohel, J.A., and Plohr, B.J., Quadratic systems of ODE's describing phenomena in shear flows of non–Newtonian fluids, Proceedings IMA Workshop on Problems that Can Change Type (Springer), accepted.

E. DiBenedetto:

DiBenedetto, E., Recent results on the Cauchy problem and initial traces for degenerate parabolic equations, Archive for Rational Mechanics and Analysis, to appear.

L.A. Peletier:

de Mottoni, P., Peletier, L.A. and Tesei, A., Free boundary eigenfunctions for a nonlinear degenerate eigenvalue problem, J. reine angew. Math. **394**, 31–58, (1989)

Springer